D0762339
OFFICIALLY WITHDRAWN

Symmetry

A Series of Books in Chemistry

Symmetry

A STEREOSCOPIC GUIDE FOR CHEMISTS

Ivan Bernal
Walter C. Hamilton
BROOKHAVEN NATIONAL LABORATORY

John S. Ricci
WINDHAM COLLEGE

W. H. Freeman and Company
SAN FRANCISCO

Printed in the United States of America

Library of Congress Catalog Card Number: 75-178258
International Standard Book Number: 0-7167-0168-5

1 2 3 4 5 6 7 8 9

To those who have instructed us in symmetry
especially
Judi and Jennifer

Tyger! Tyger! burning bright
In the forests of the night,
What immortal hand or eye
Dare frame thy fearful symmetry?

WILLIAM BLAKE

Preface

The geometrical arrangement of atoms in a molecule influences in large part how that molecule behaves in its interactions with other molecules. Such influences are responsible for the physical properties of a single substance in any state of aggregation. They are also responsible for the details of the reactions of molecules with one another to produce new molecules—chemistry. The student of chemistry can understand his subject only if he has a deep understanding of geometrical structure. How are the atoms connected to one another? How far apart are they? What shapes do the molecules have?

Central to the description of molecules and their properties is the concept of symmetry, by which the pictorial and mathematical descriptions of molecules can be simplified. In this book we introduce the concept of symmetry operations and groups of operations and present—for the most part in three-dimensional pictures—the way in which these operations can be used in the description of molecular shapes. The student who gains a familiarity with the appearance of various symmetry properties in three dimensions should be adept at immediately classifying the symmetry of a new and unfamiliar molecule.

We discuss symmetry in general terms in Section I and draw attention to its pervasiveness in history, art, and science. In Section II we introduce the elementary symmetry operations and discuss the concepts of groups of symmetry operations. In Section III, we enumerate the two-dimensional crystallographic point groups. The heart of the book is Section IV.

There we present the three-dimensional crystallographic point groups together with a number of other three-dimensional point groups useful in molecular structure descriptions. This section is illustrated with three-dimensional stereoscopic drawings.

The book was planned for students with a background in secondary school mathematics and chemistry. As such, it is a useful supplementary textbook for a first-year college course in chemistry or a secondary school honors course. Some of the chemical concepts and terminology that may be unfamiliar to the beginning student are explained in the glossary that follows the text.

We would like to express our appreciation to the staff of the Photography and Graphic Arts Division and the Central Scientific Computing Facility at the Brookhaven National Laboratory for their continual cooperation in the production of the three-dimensional drawings, to Dr. C. K. Johnson for making available his computer program ORTEP, and to Mrs. Bonnie Wesolowski for her careful preparation of the typescript.

Upton, New York
Putney, Vermont
January 1972

Ivan Bernal
Walter C. Hamilton
John S. Ricci

Contents

Some Practical Advice to the Student

The development of an understanding of symmetry is essential to the full appreciation of molecular structure, and we believe thorough study of the text and the illustrations in this book can aid in that development. We think the following information about this book can increase the usefulness of the book and your enjoyment of it.

The Stereoscopic Illustrations

The complexities of molecular architecture in three dimensions determine the most interesting properties of molecules. The geometry of a complex molecule is best understood by holding a three-dimensional ball-and-stick model in the hands. If such a model is not available, the next best thing is a stereoscopic drawing. The perception of the three-dimensionality of an object results largely from the fact that a slightly different image is presented to the left eye than to the right eye because the eyes are a short distance apart. The mind has learned to interpret this difference as a measurement of the relative distance of objects from the viewer. The illusion of depth can be created with two-dimensional images if two slightly differing images are viewed together, one with the left eye and one with the right eye. Such plane figures may be prepared by actual photography of an object from two points separated in location by about the same distance that separates the two human eyes. Alternatively, the proper positions of objects in the two figures may be computed: the drawings in this book were prepared with the help of an elegant computer program, ORTEP, written by Carroll Johnson of the Oak Ridge National Laboratory.

A viewer is provided with this book to aid in viewing the stereo drawings in the section on three-dimensional point groups. The viewer should be stood on the page so each lens is over half of each stereo drawing, so the right eye sees only the right image and the left eye only the left image. The images will then merge into a single three-dimensional picture. If both of your eyes are normal, you should obtain the stereoscopic effect with little trouble, although a little practice may be needed. Tilting your head slightly and starting with your eyes very close to the drawing may help. After you have once done it, it is almost impossible not to let the images merge.

Many people can obtain the stereoscopic effect without the use of any supplementary optical device. Hold the drawing about ten inches from your eyes and relax them as if staring into the distance. The image of the right-hand member of the stereo pair viewed with the right eye and the image of the left-hand member viewed with the left eye will move together to produce an apparent three-dimensional object. Ghosts—the right-hand object viewed by the left eye and the left-hand object viewed by the right-eye—surround the merged figure and may be distracting. These may be eliminated by holding a pad of paper half way between the two drawings and extending it to your nose.

The Choice of Illustrations

In illustrating three-dimensional symmetry groups, it seemed desirable to choose examples that could also give the student a feeling for the variety of molecular structures that are possible, moreover, to teach him a little chemistry. When we began searching the scientific literature for authentic chemical illustrations for this book we decided to use the following guidelines: (*a*) the molecule should have precisely the desired symmetry; (*b*) it should have some inherent chemical importance or have been used for an interesting purpose; (*c*) it should be relatively simple in chemical composition; and (*d*) it should be a tractable representational model. By this we mean that the atoms are so situated with respect to one another that the symmetry of the molecule can effectively be displayed with a minimum number of overlapping atoms along the observer's line of sight.

To obtain recent examples that satisfied these guidelines, we conducted an extensive search of the literature of molecular structures for the period 1960–1970. In some instances we used older studies to supplement the details we found in more recent standard sources. It was our frustrating experience to find an overabundance of examples of certain point groups and few or none of others. Therefore, we relaxed our requirements and improvised a few molecules whose structures have not yet been determined but that could belong in theory to the point group in question. For some illustrations, we accepted an attractive molecule whose idealized symmetry was what we desired even if in nature the structure presented minor deviations from exact compliance with the symmetry elements of the point group. This was often true of molecules found in crystals, where packing forces might cause the molecular structure to depart from being ideally symmetric. For example, the molecule hexamethylenetetramine, $C_6H_{12}N_4$, is unusual in having such high symmetry (cubic) for an organic compound. Although the molecule does have perfect $\bar{4}3m$ symmetry in crystals of the pure substance, a chemical derivative—the hexahydrate—crystallizes in such a way as to distort the molecule to a lower symmetry. A highly symmetrical molecule such as hexamethylenetetramine provides an excellent example of the usefulness of stereoscopic drawings. Compare the flat representation, which is found in many organic chemistry textbooks,

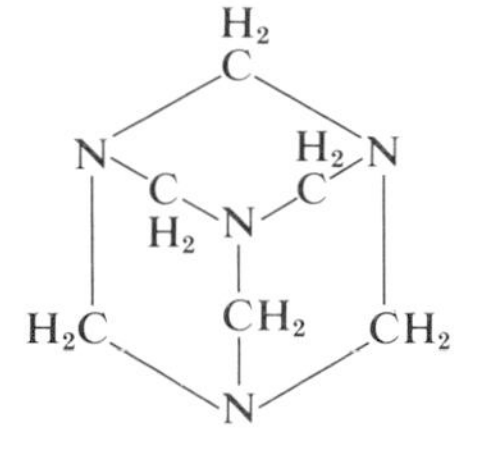

with the following stereoscopic drawing.

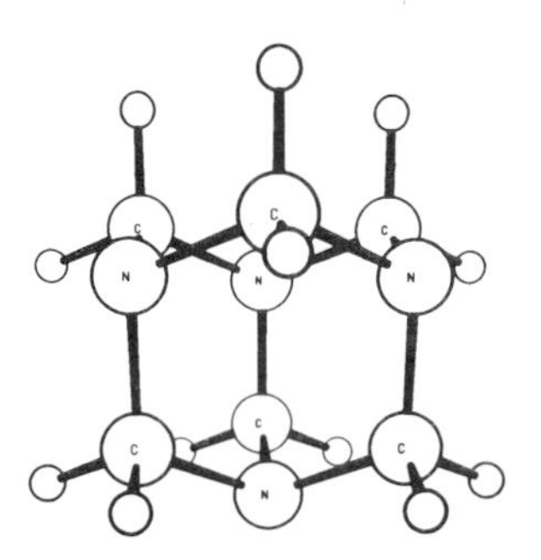
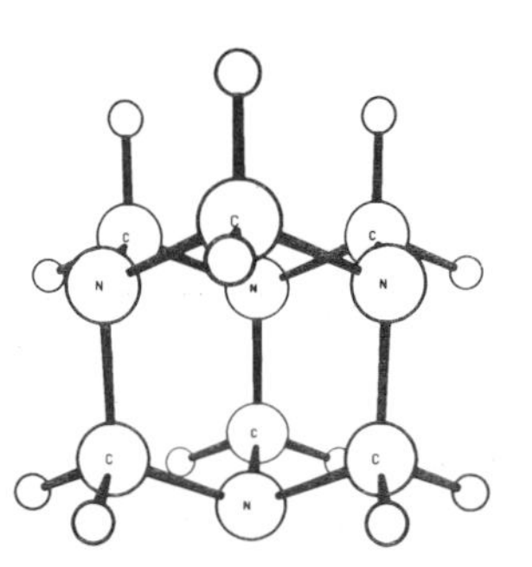

Presentation of the Illustrations

In many discussions of symmetry, it has been conventional to present the drawings in such a way that the major symmetry axis is vertical and the molecule is viewed from above. This is probably a useful convention for two-dimensional drawings and one the student should become accustomed to. In most of our drawings, however, we have chosen to present molecules in an arbitrary, although pleasing, orientation such that the principal symmetry axis may not be vertical. We believe that it is good practice for the student to recognize the symmetry elements of a molecule regardless of the aspect in which the molecule is presented to him. For each molecule, he should be sure that he has correctly identified the symmetry elements of the point group in question.

The Text for the Illustrations

Each person or group of persons who studied one of the molecules shown in this book had a specific reason for wanting to do so. To convey to the reader the broad range of ideas and reasons that motivate a scientist to carry out an accurate structural study, in conjunction with each stereo illustration in our book we have paraphrased some of the words of original papers in which scientists explained their investigations. We hope that in doing this we have preserved some of the flavor of each researcher's original writing. In most cases we also provided some commentary and descriptive material of our own. However, we urge the student not to end his reading just because he has reached a reference at the foot of one of our pages, but to go further and to consult the original sources: a good cross section of the literature in molecular structure investigation awaits him.

The names of the molecules in the illustrations are those given by the authors of the original papers; not all of these names conform with the rules that have been established by committees of the International Union of Pure and Applied Chemistry. It is beyond the scope of this book to include a comprehensive discussion of chemical nomenclature. The interested student may consult appropriate advanced textbooks or a handbook such as *The Handbook of Physics and Chemistry*, 49th Edition (Chemical Rubber Publishing Co., Cleveland, 1968), which has useful, compact discussions of the nomenclature of organic (page C-1) and inorganic (page B-148) compounds.

The Problems

Beginning on page 157 are a number of illustrated problems. In each, you will be asked to give the point group symmetry of the molecule or crystal structure of the example. The answers are not always obvious. It will probably be extremely helpful to you to discuss these problems with your teacher and fellow students. These exercises should serve as a stimulus to you in reading the chemical literature: every time you meet a new molecule, ask yourself, "What is the symmetry of this molecule?" We hope your interest in symmetry will extend further. The next time you are in a house with papered walls, look at the design of the wall-paper: it is often a classic illustration of the use of plane groups. So, look for the symmetry elements hidden in the roses (or whatever the motif) and determine the point group.

Glossary of Terms

Although we do not discuss chemical nomenclature in detail, often in the discussion of the chemical examples of the symmetry groups we do use words that may not be entirely familiar to all who use this book. Therefore we have provided a Glossary (page 173) that may be helpful to the reader. Each term included in the glossary is denoted with bold face type in the text.

Further Reading

The serious student of symmetry and the geometrical aspects of molecular structure is urged to delve into some of the books described in the Bibliography, which begins on page 179.

I

INTRODUCTION

Symmetry, Art, and Science

In the daily interactions with his environment and with his peers, man tends either to overlook or to forget the obvious and the commonplace. Thus it is with symmetry, which pervades our activities, pleasures, and tools, which determines our health and behavior, and which to a marked degree controls our biological existence. Although our chief concern in this book will be with the use of symmetry concepts in the description of chemical compounds, the student should be aware that these concepts have pervaded many fields of endeavor since the earliest beginnings of human thought. Widely disparate intellectual pursuits are often affected by the same concepts in symmetry.

What is symmetry? The word is constructed of the Greek root *metron*, to measure, and the prefix *syn* (becoming *sym* before the letter *m*), along or together. To measure together? Two or more aspects of a symmetric figure do have the same measure. Symmetry is the characteristic of a pattern or object that leads us to say two or more parts of it are in some respects the same: this part is like that part.

In the world around us, plants and animals usually exhibit symmetry in their external forms. The most obvious example is man himself. Divide a man down the middle and there is a sense in which both halves are the same. Like most vertebrates, he has a left side and a right side that are functionally and visually nearly identical. The left foot looks like the right foot and both perform the same function. Put a man, or a horse, in front of a mirror and the image still looks like a perfectly reasonable man or horse. This is not true of all of man's symbols—say, of all letters of the English alphabet—although it is for some.

But for that matter man himself rarely has perfect left–right symmetry. His left side does not look exactly like his right side, nor does it always perform identically. Nevertheless, the approximate symmetry is always

A comparison of symmetry and assymetry in man.
(Photo of Edgar Allan Poe at left:
Courtesy American Antiquarian Society.)

there, and we perceive the essential equality of left and right despite small differences. If the left and right halves are markedly different, that man or animal is considered remarkable: consider the perverse appearance of the flounder. Recognizing approximate symmetry is useful in chemistry, as it is in zoology, for even though the symmetry may fall short of perfection, that there is near-symmetry is usually important in determining behavior.

Many lower animals have the left–right mirror symmetry of a man but sometimes exhibit other types of symmetry too. The starfish, for example, has five identical arms.

Among plants, we again observe the mirror symmetry in most leaves. The idealized clover exhibits additional types of symmetry in its three (or four) identical leaves.

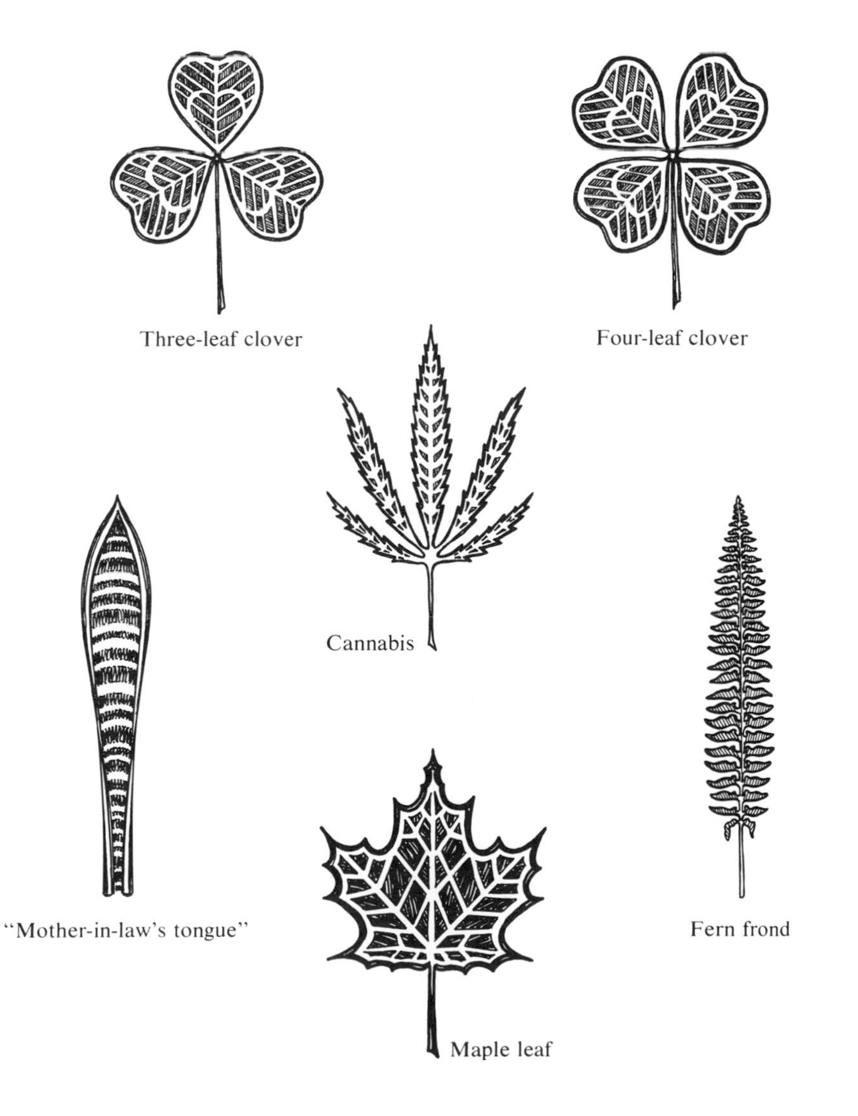

The leaves of the sassafras tree are surprising, however, because some possess mirror symmetry but most appear to be the shape of either a right-hand or left-hand mitten.

A freestanding fir tree has approximate cylindrical symmetry. Rotate it around the trunk and its approximate symmetry remains the same.

An example of rotational symmetry

In objects man has created, symmetrical proportioning is one of the oldest of concepts. Among the earliest objects devised, woven containers for gathering and storing nuts, fruit, and other edibles were made durable and strong through symmetrical construction. Many of these functional containers are still manufactured today in primitive societies, fine examples of symmetrical arrangements of straw, palm leaves, and bark, often little changed since prehistory.

Obviously, if these beautiful patterns of symmetrically woven fibers had failed to provide strength and durability, they would have been rejected, to be replaced with containers of more practical design. Thus, man found early that symmetry and function are often linked together.

An African (Bushongo) raffia pattern exhibiting —with minor distortions by the artist—plane symmetry $2mm$. (Compare with page 47.)

Symmetry as an artistic concept and a medium of expression has been with us since man's earliest attempts at communication. For example, one of the earliest man-made wall paintings, found at Çatalhüyük, Asia Minor, depicts a ritual leopard dance.

Ritual leopard dance. Fresco on the wall of a shrine at Çatalhüyük, Asia Minor. This early sixth millenium B.C. wall painting is among the earliest known. (From M. Grant, *The Ancient Mediterranean,* New York, Charles Scribner's Sons, 1969.)

At first view the tribesmen in the painting may seem to show only the usual symmetry of the human form. But notice the dancer in the center of the scene. His body is painted half black and half white; by juxtaposing these two colors, the artist has transformed the inherent human symmetry of right and left halves into something more complex and uniquely beautiful. Although we will not treat two-color symmetry we will remark that the concept of two-color symmetry is of major importance in the study of bipolarity: in magnets, for example, the two colors can be interpreted as north and south poles. The use of color as a symmetry operation greatly increases the number of possible symmetry groups; the 1651 two-color symmetry groups in three dimensions have been tabulated by V. A. Koptsik (see Bibliography).

As man became more sedentary and could spare time to appreciate beauty in his daily life he began purposely to use symmetrical patterns because of the visual pleasure they could add to his practical constructs. Some of the sheer visual pleasure to be derived from symmetrical patterns can be sampled by studying this masterpiece of Peruvian weaving.

A cotton and wool tapestry from the central coast of Peru, woven sometime between 1000 and 1500 A.D. (From C. C. Mayer, *Masterpieces of Western Textiles,* Chicago, The Art Institute of Chicago, 1969.)

If we concentrate on the pattern generated by the playful black monkeys, we see that there is an approximate 2-fold symmetry axis relating not only adjacent monkeys but the entire pattern of the blanket. It is interesting to note that the artist gave himself the pleasure of breaking perfect 2-fold symmetry for the entire pattern by putting masks on some of the light colored figures but not on others. Possibly the artist found it desirable to be slightly inconsistent mathematically. With the exception of this detail, each figure is related to many others by 180° rotations and by simple translation.

Examples of symmetry in architecture stand everywhere. One of the examples made interesting for us because it is so purely geometric is provided by the work of Buckminster Fuller, who has achieved considerable renown through his use of lightweight frames of high three-dimensional symmetry over which he stretches the building's outer skin.

United States Pavilion at Expo 67. Montreal, designed by R. Buckminster Fuller.

Mr. Fuller's imagination along these lines is comparable to the masterful way in which the elements beryllium or boron arrange themselves in polyhedral arrays. On the next page are some examples of the polyhedral configurations that beryllium adopts in a crystalline alloy with rhodium. The beautiful polyhedra in the molecular structures of boron compounds have been well illustrated by W. N. Lipscomb and by E. L. Muetterties and W. H. Knoth (see Bibliography).

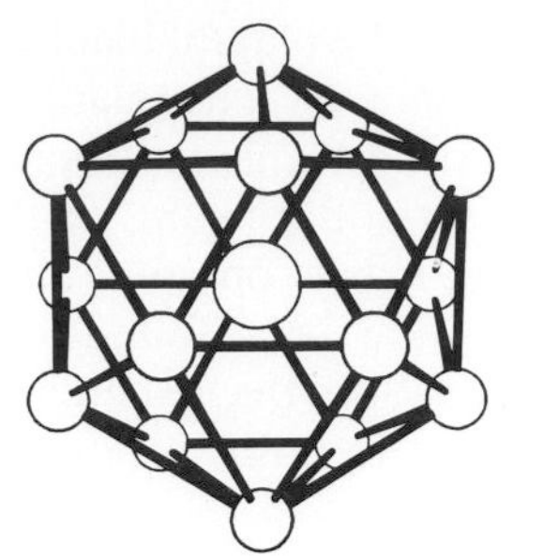
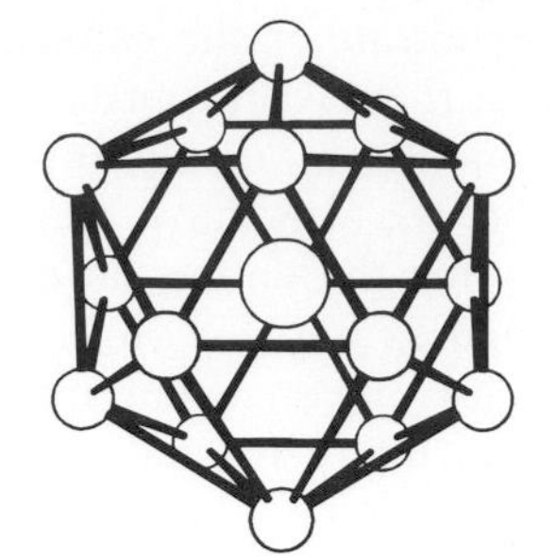

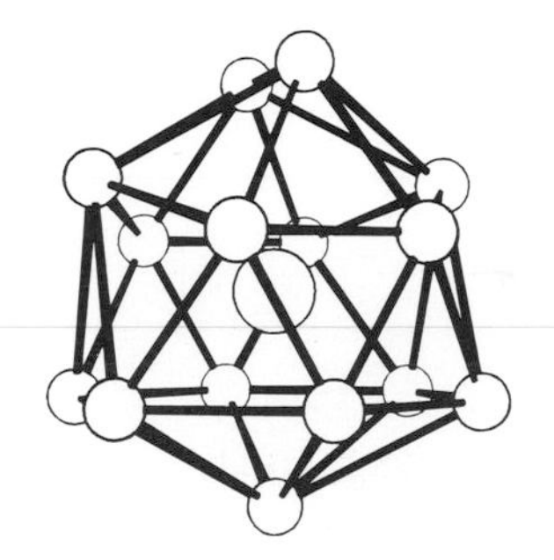
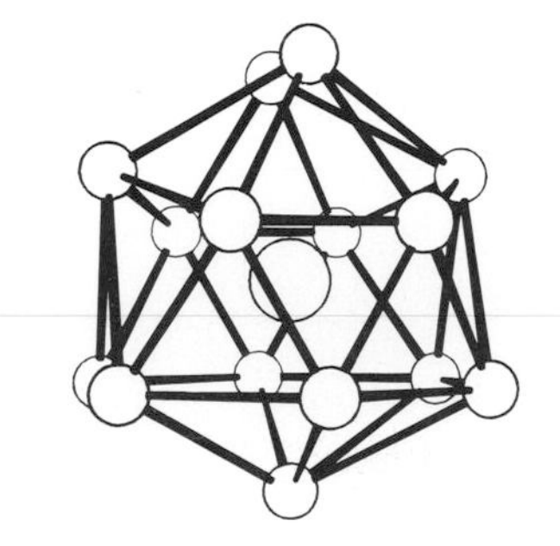

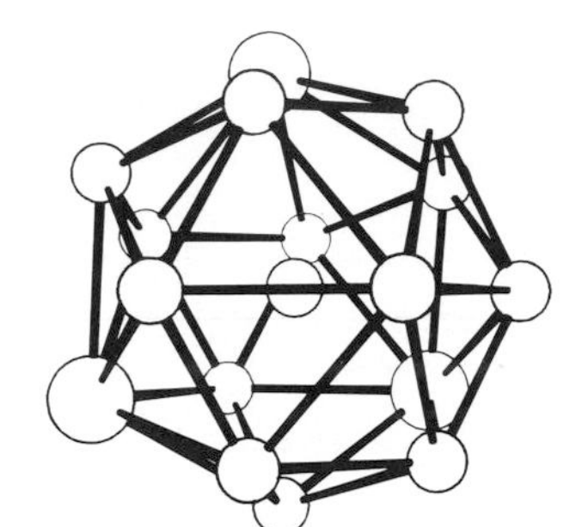
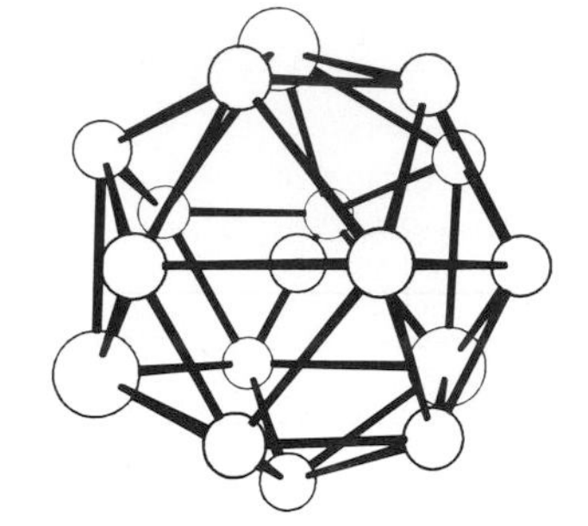

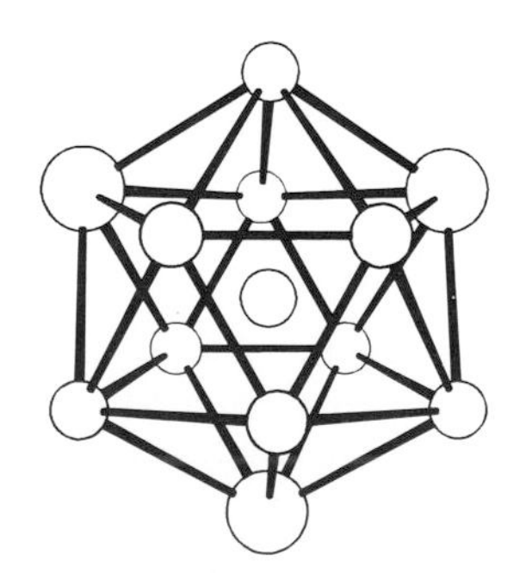
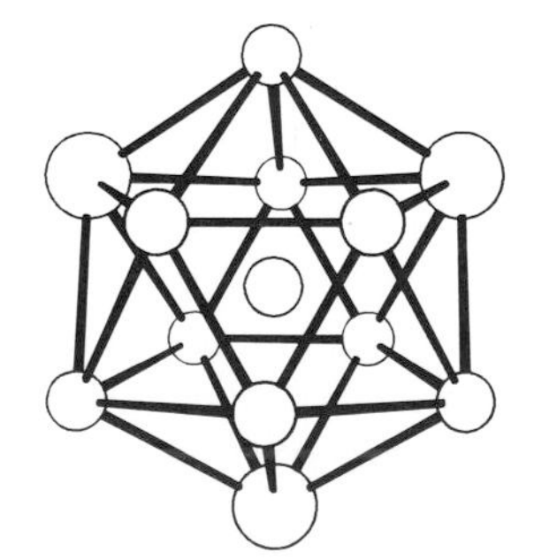

Coordination polyhedra of atoms in the structure of an alloy, $RhBe_{6.6}$. (Adapted from Q. Johnson, G. S. Smith, O. H. Krikorian, and D. E. Sands, *Acta Crystallographica*, **B**26, 109, 1970.)

Water, the life-giving fluid, is also a versatile builder of symmetric structures; these are sometimes useful but are also sometimes destructive. The ability of water to solidify at temperatures above its normal freezing point while encaging an impurity is sometimes responsible for fractures in oil pipelines and for the destruction, in the plains regions of the Midwest, of cells within grain such as wheat and corn when the temperature falls suddenly. The compounds that result from the incarceration of impurity molecules in cages of another molecule are known as clathrates, and their symmetry, beauty, and significance are remarkable.

A clathrate hydrate in which many water molecules form a cage surrounding a tetra-alkyl ammonium ion in a symmetrical polyhedral arrangement. (Courtesy J. F. Catchpool.)

It is of interest that the formation of such water cages has been suggested as being responsible for the ability of substances such as chloroform to produce anesthesia (L. Pauling, *Science*, **134**, 15 (1961)).

It seems perhaps incongruous to put architecture, the freezing of oil pipes, the frost damage to midwestern grain crops, alloys, anesthesia, and crystallography in the same bag. The fact, is, however, that all these topics come together at the juncture of molecular structure and symmetry and any fundamental finding or contribution from one is bound in the long run to affect the other, if one investigator is sufficiently curious to look into another's backyard.

The relationship between the left and right hands is at the very heart of the juncture of structure and symmetry; it is one of the most basic symmetry concepts. Two objects, identical in shape but not superimposable because one is right and one is left are called *enantiomorphs*. This right and left handedness is of prime importance to our physiological well-being and even to the functioning of life itself. The molecular building blocks from which living organisms form the macromolecules that constitute most biological structures and that transmit hereditary characteristics exist in left- and right-handed forms, of which only one is generally used.

Molecules of DNA, the basic material of the gene, consist of two right-handed chains that coil about each other in a helical form. All but one of the amino acids in proteins are enantiomorphic—that is, all but glycine naturally have left-handed and right-handed forms; but inexplicably, proteins present in plants and animals are made up (with rare exceptions) of only left-handed amino acids. An interesting illustration of the selectivity of bodily processes for either the right-handed (dextro) or left-handed (levo) forms has been succinctly described by N. Sharon (*Scientific American*, **220**, 92 (May, 1969)):

> Only a limited number of the twenty **amino acids** that form **protein** are present in the cell-wall material. In the walls of many bacteria, the amino acids are glutamic acid, glycine, lysine and alanine. Two of the amino acids, however, are found in an "unnatural" configuration, that is, certain of the atoms or groups of atoms in their molecules are arranged differently from the way they are in the amino acids of other natural substances. This configuration, which is designated dextro (D), is a mirror image of the "normal" configuration, levo (L). . . . The amino acids of the cell wall are linked by **peptide** bonds (HN—CO), but they are not susceptible to digestion by **enzymes**, such as trypsin or pepsin, that act to split such bonds (proteolytic enzymes). This resistance may well be related to the unusual D configuration of the wall components glutamic acid and alanine.

Thus, in their struggle for survival, bacteria protect themselves with a mirror operation.

Left-or-right orientation is also true of many drugs: morphine, for instance, exists in two enantiomorphic forms, but only the left-handed form of the drug is active; the right-handed form is neither a useful pain killer nor is it addictive. Both dextro and levo amphetamines are physiologically active; however, the former is as much as ten times more active in such biological functions as the enhancement of locomotor activity (K. M. Taylor and S. H. Snyder, *Science*, **168**, 1847 (1970)). The popular and controversial natural drug marijuana consists of a mixture, of which most or all active components can be classified as either right or left handed.

A significant property of right and left handedness, one which will be referred to often in subsequent sections, is optical activity. A substance in which molecules exist in either right- or left-handed forms is capable of rotating the plane of polarization of a polarized light beam passing through it; this is known as **optical activity**.

The Representation of Symmetry

Our approach to the study of symmetry in this book is the visual approach. What happens to the appearance of an object when symmetry operations are applied to it? What is the resulting appearance of an assemblage of symmetry-related objects? In order to fully exploit the possibilities of this approach, it is necessary to choose as a basic element of the illustrations an object that has no inherent symmetry of its own in the space (usually two- or three-dimensional) of the illustration. For example, if one wishes to operate in two-dimensional space, as do artists who design wallpaper, suitable unconstrained asymmetric objects are the foot, the hand, and the horse.

In three-dimensional space, however, the horse (like the man) has the problem that he has a built-in symmetry element. The approximate mirror plane of the horse converts the left half of the horse into the right half. A horse in a mirror still looks like a horse—indistinguishable from the original—and as such it is unsuitable for dramatic illustration of the mirror operation. We assume that the horse does not have a large distinguishable spot on either the left or right side. If the horse has spots, the mirror image can certainly be distinguished, just as the mirror image of a foot or hand can easily be distinguished from a left and right foot or hand by referring to a big toe or thumb.

We have, as a matter of fact, chosen the foot as the asymmetric object for our two-dimensional illustrations. In three dimensions it seemed preferable to use another object, which satisfied the following criteria: (*a*) it should be asymmetric; (*b*) it should be of simple nature, to avoid cluttering the drawings; (*c*) it should be readily describable in terms of a small number of three-dimensional coordinates x, y, z so that the preparation of the drawings could be easily programmed on a digital computer. A figure that satisfies these criteria is shown here.

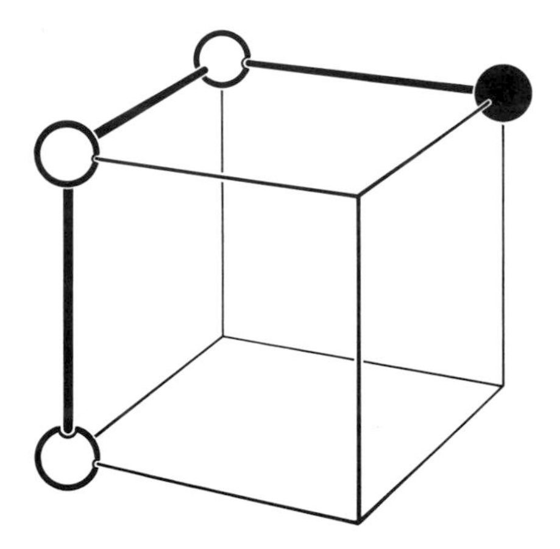

Consisting of four balls on adjacent corners of a cube, this very simple figure has no symmetry whatever provided any one of the four balls is made different from the other three. It is the basic unit (or asymmetric unit) of design we have used throughout Section IV on three-dimensional symmetry and is interesting in its own right since it contains the minimum number of points (four) necessary to make it a three-dimensional figure. It also provides a simple illustration of a principle that is of great importance in chemistry and in biology. Consider the effect of reflecting the object in a mirror that is parallel to the lines 3–4 and 1–2. The result is shown on the next page.

In his mind a reader should try to rotate and manipulate the cube in the left half of the figure so that it is coincident with that in the right half of the figure. After some time—how much will depend on his tenacity and endurance—the reader will be convinced that the goal cannot be achieved short of reversing the original mirror operation. This would be true even

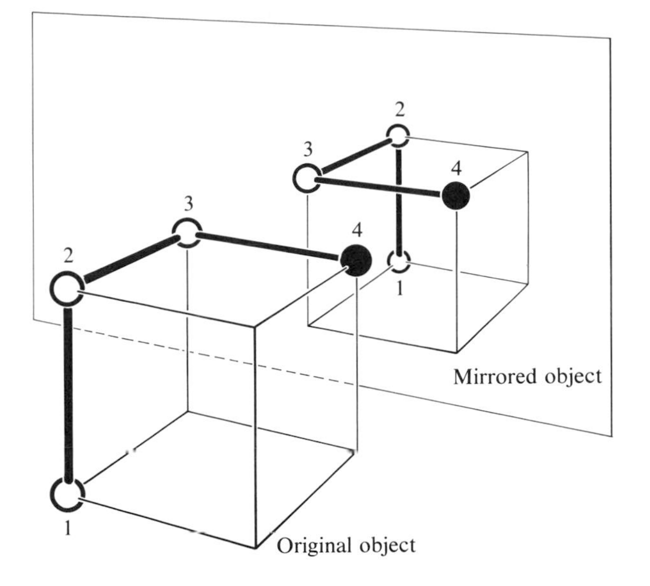

if the balls were identical. However, identical balls would form an object with a 2-fold axis—one still not identical with its mirror image but nevertheless possessing some symmetry. In the previous chapter, we spoke of molecules that exist in both left- and right-handed forms—differing in that they are mirror images of one another. The two distinguishable objects in the figure above provide a simple example of this property.

II

BASIC CONCEPTS

The Concept of a Symmetry Operation

Consider the patterns represented in the three parts of the figure below. What are the characteristics of these patterns that lead us to say that the patterns in part *a* have no symmetry but those in parts *b* and *c* do? We might even say that the patterns in *c* have more symmetry than those in *b*.

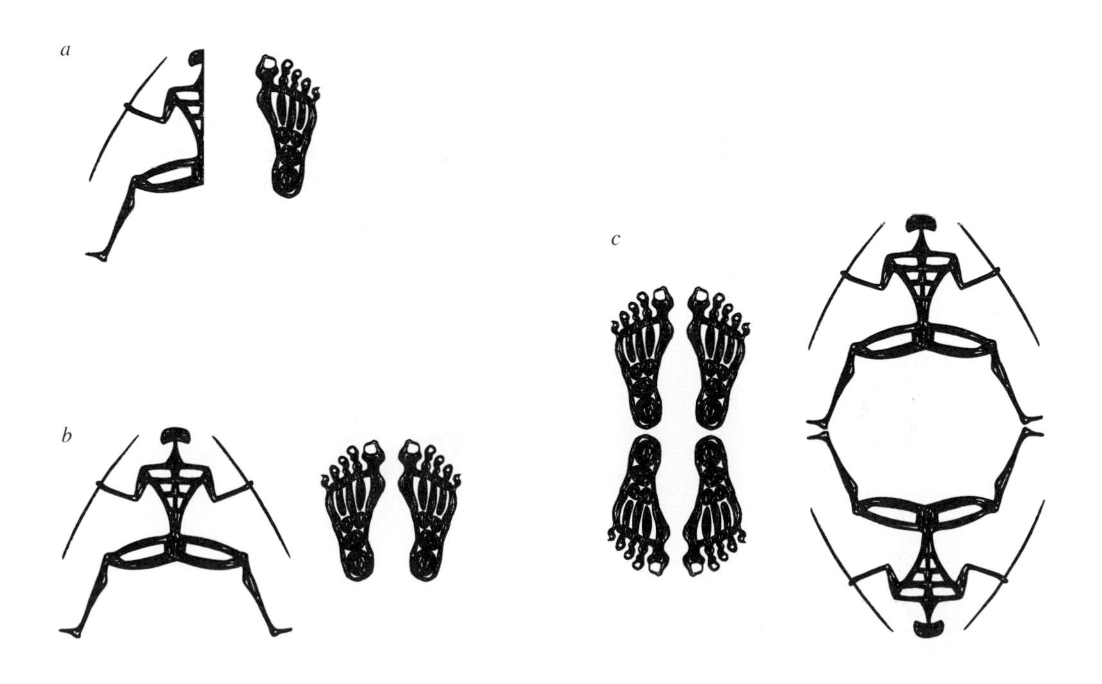

Let us look at the silhouette figure shown at *b* on the previous page and label the legs A and B for the moment.

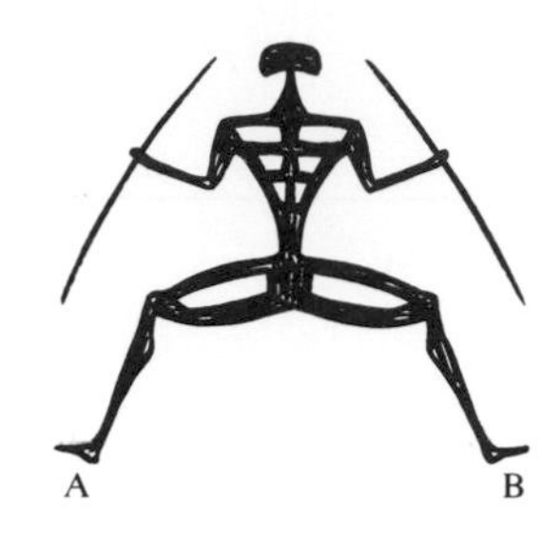

Now let us rotate the figure around the vertical line by 180° (half a complete rotation).

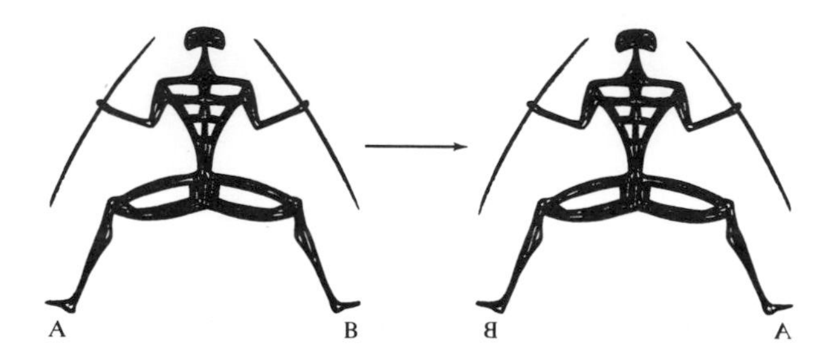

Because the legs are identified by letters, the two figures can be differentiated. But remove the letters and the figures are identical.

The operation of rotation by 180° has left the appearance of the figure unchanged. Such an operation, which leaves the appearance of a figure unchanged after some real interchange of points, is called a symmetry operation. The figure is said to be symmetric.

A symmetry operation may be described in terms of the rule by which any point in the figure is transformed into another such point. In the figure below, a point with coordinates x, y goes into another point with coordinates $-x$, y when the operation is carried out. Points that are unchanged, in this example those with coordinates 0, y, are said to lie on the symmetry element.

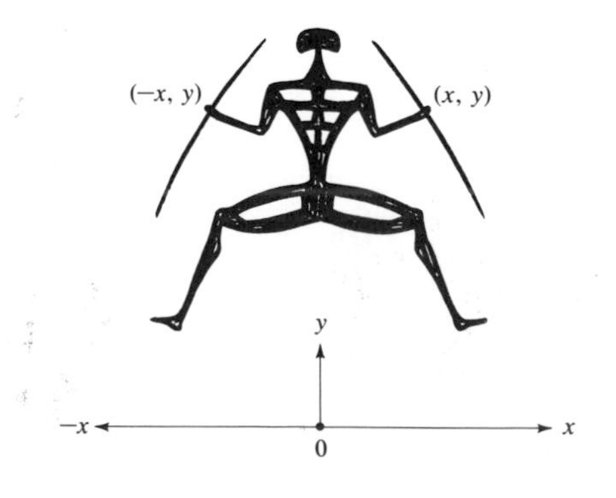

Consider again the right foot shown at the beginning of this chapter:

Imagine that a mirror is placed adjacent to this foot and is slowly translated from left to right. You may find it useful actually to do this.

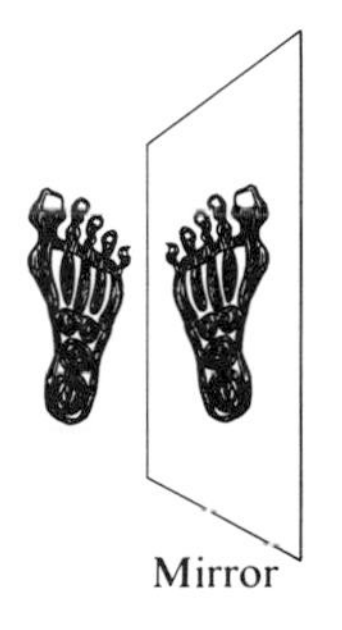

An image of the foot is produced in the mirror, but the image is a left foot, not a right foot. The two feet together form a *symmetric* figure in that there is a one-to-one correspondence between any point on the right foot and a similar point on the left foot. The asymmetric unit of the figure—either the right foot or the left foot—is reproduced by the *mirror* symmetry operation to produce a foot of the opposite handedness. Perhaps we should speak of footedness, but objects of all kinds that bear a mirror relation to one another are referred to as left- and right-handed objects. A more sophisticated term for this property is **chirality**. Your left and right feet have opposite chirality.

Another type of symmetry is illustrated by the following figure.

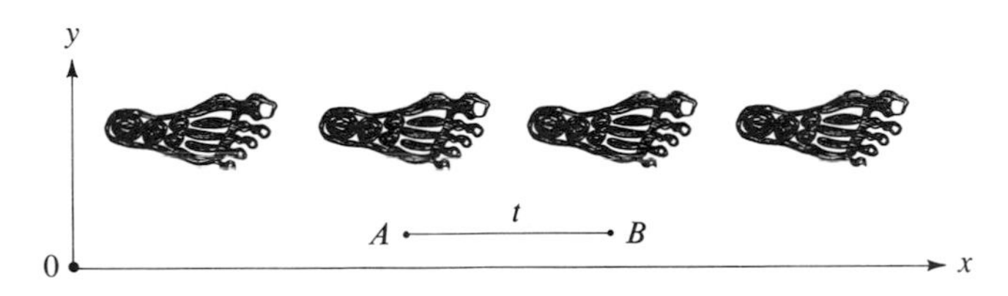

All right feet, all going the same way; to every point on one corresponds a point on the next. The transformation relating two successive feet A and B may be represented as

$$x_B = x_A + t$$

$$y_B = y_A$$

This symmetry is called pure translational symmetry. For every foot there is a following foot, and if every foot is identical the chain of footprints must be infinite.

In our present discussions of the symmetry of molecules we will be mainly concerned with symmetry operations that do not involve translations, namely those for which at least one point in the figure does not change position as a result of the symmetry operation. Such symmetry, which leaves at least one point unmoved, is known as point symmetry.

Symmetry Operations in Three Dimensions

A symmetry operation S is a coordinate transformation that takes a point x, y, z and converts it to another point x', y', z'. This may be denoted by

$$S: \quad x, y, z \rightarrow x', y', z'$$

For example, a mirror operation in three dimensions (with the mirror plane perpendicular to the y axis and passing through the origin) takes a point x, y, z and transforms it into the point $x, -y, z$.

$$m: \quad x, y, z \rightarrow x, -y, z$$

In visualizing this operation, the reader may find useful the construction indicated in the figure on the next page.

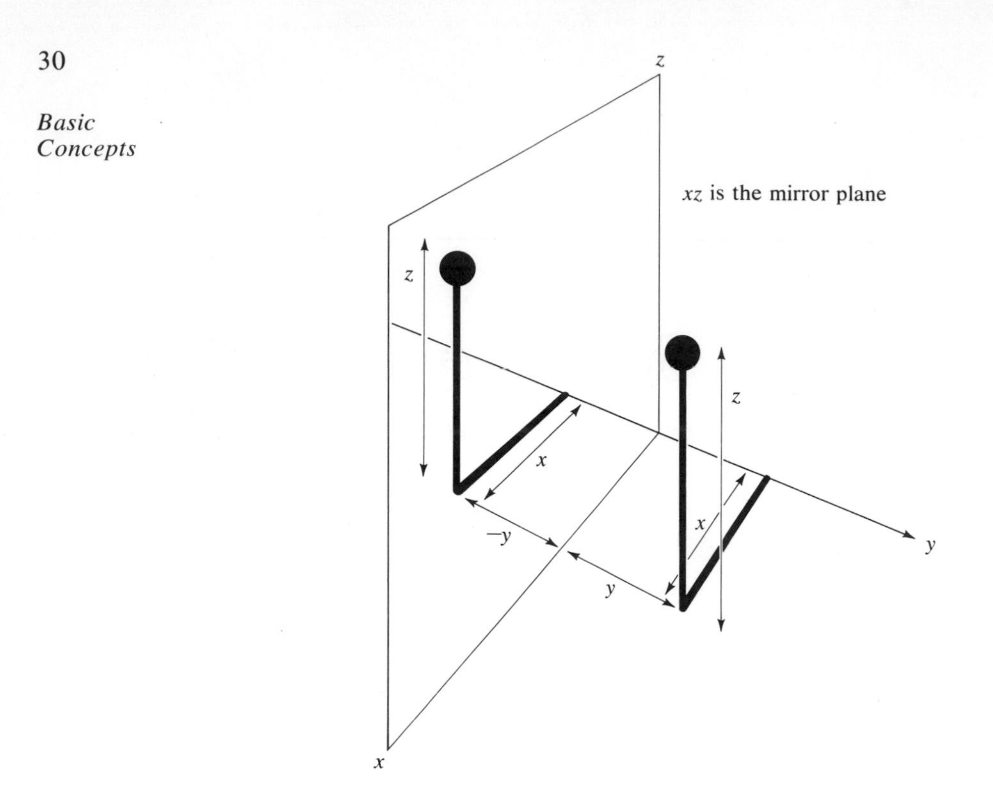

It is useful to consider that the operation that does nothing to a figure, leaving all points fixed,

$$E: \quad x, y, z \rightarrow x, y, z$$

is also a symmetry operation, called for obvious reasons the *identity operation*.

Groups of Symmetry Operations

In any symmetrical figure, repetition of symmetry operations of the same or of different kinds results in figures that are indistinguishable from the original; thus, a sequence of symmetry operations must be equivalent to a single symmetry operation. For example, a mirror operation m_1 followed by a mirror operation m_2 has the following result

$$m_1: \quad x, y, z \rightarrow -x, y, z$$

$$m_2: \quad -x, y, z \rightarrow x, y, z$$

The combination of the two operations may be written as

$$m_2 m_1: \quad x, y, z \rightarrow x, y, z$$

where we adopt the usual convention that we operate from right to left in applying symmetry operations; thus

$$ABC$$

means

first do C
then do B
then do A

Since the identity operation gives the same result:

$$E: \quad x, y, z \rightarrow x, y, z$$

we may write

$$m_2 m_1 = E$$

This means "The result of applying the two mirror planes in sequence is the same as the application of the identity operation." The equation looks like an ordinary algebraic product. We may by analogy make the statement that the *product* of the two symmetry operations m_1 and m_2 is the symmetry operation E. It would be good practice for the reader to illustrate the conclusions made in this paragraph with the use of drawings such as that on page 30.

A figure described by the set of symmetry operations, together with an assymetric unit, for example, our

provides us with an illustration of a mathematical *group*. The following properties suffice to determine a symmetry group: (*a*) the product of any two symmetry operations is also a symmetry operation of the group; (*b*) there is an identity symmetry operation; and (*c*) every operation has an inverse operation also in the group. The description of a particular symmetry group may be expressed in terms of a group multiplication table, which gives the relationships of various symmetry elements to one another. The members of the group – the symmetry operations – are usually referred to as the *elements* of the group.

Thus in a symmetry group with the two elements E and m, we may write

$$EE = E$$
$$Em = m$$
$$mE = m$$
$$mm = E$$

or

	E	m
E	E	m
m	m	E

or even more simply

$$\begin{matrix} E & m \\ m & E \end{matrix}$$

A group of symmetry operations that leaves at least one point invariant (in the same place) is known as a point group, and it is the study of point groups that is of primary importance in molecular symmetry.

Point Group Symmetry in the Plane

Let us first consider objects confined to two dimensions, points of the object being described by two coordinates*

$$x, y$$

A symmetry operation S transforms a point x, y as follows:

$$S: \quad x, y \rightarrow x', y'$$

There are two types of operations that leave at least one point invariant.

*In most of our drawings we will conventionally take the x axis as directed toward the right and the y axis toward the top of the page. In three dimensions the z axis will be perpendicular to the plane of the paper and directed toward the reader. This convention defines a right-handed coordinate system.

One of these types we have already treated, the mirror operation†

$$m_y:\quad x, y \rightarrow -x, y$$

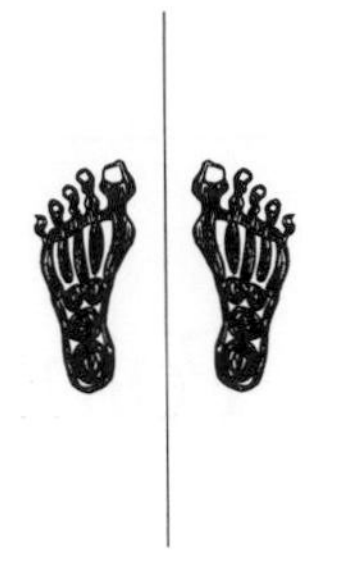

or

$$m_x:\quad x, y \rightarrow x, -y$$

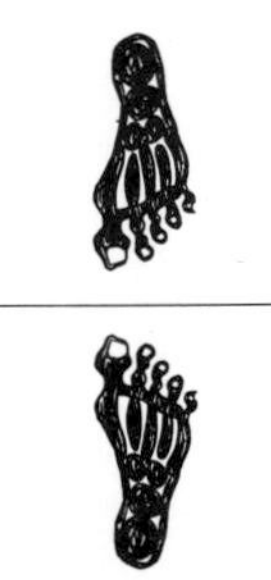

The mirror operation leaves an entire line of points invariant; this line is the mirror line. (We may occasionally use the term mirror *plane*, even in two dimensions.) A mirror always converts a right-handed to a left-handed object.

The second type of operation is the rotation axis.

$$n:\quad x, y \rightarrow \left(\cos\frac{2\pi}{n}\right)x - \left(\sin\frac{2\pi}{n}\right)y,\ \left(\sin\frac{2\pi}{n}\right)x + \left(\cos\frac{2\pi}{n}\right)y$$

where n is an integer. Thus, if n is 2, we obtain

$$2:\quad x, y \rightarrow -x, -y$$

†For the two-dimensional groups we have taken m_x and m_y to be the mirror lines *parallel* to the x and y axes.

Only certain combinations of these basic elements lead to groups that contain a finite number of symmetry operations.

As examples, let us consider point groups in the plane that contain rotation axes up to order 6. Of these, the point groups of order 5 are unique in that they cannot generate space-filling figures in the plane. Consider the problem of filling the plane—laying a tile floor, for example—with identical regular polygons of order n. The exterior angle at each vertex is $(180 - 360/n)°$. An integral number v of such vertices must fit together to add up to 360°. Thus

$$\left(180 - \frac{360}{n}\right)v = 360$$

where n and v are integers. This equation has solutions only for the pairs $(n, v) = (3, 6), (4, 4)$, and $(6, 3)$. Regular pentagons cannot fill the plane. The point groups with rotation axes of orders 1, 2, 3, 4, and 6 can form plane-filling figures and they are known as the crystallographic plane point groups. Point groups are usually named in such a way that the name or *symbol* gives enough information about the group that all group elements may be generated.

Thus, consider the noncrystallographic point group that is generated by a 5-fold axis. The operations* of the group are

(1) A rotation by 72°, denoted 5
(2) A rotation by 144°, denoted 5^2
(3) A rotation by 216°, denoted 5^3
(4) A rotation by 288°, denoted 5^4
(5) A rotation by 360°, denoted $5^5 = E$, the identity element.

The name of this group is simply '5', since all elements are generated by the single 5-fold rotation axis. The *order* of the group—the number of different symmetry operations—is also 5.

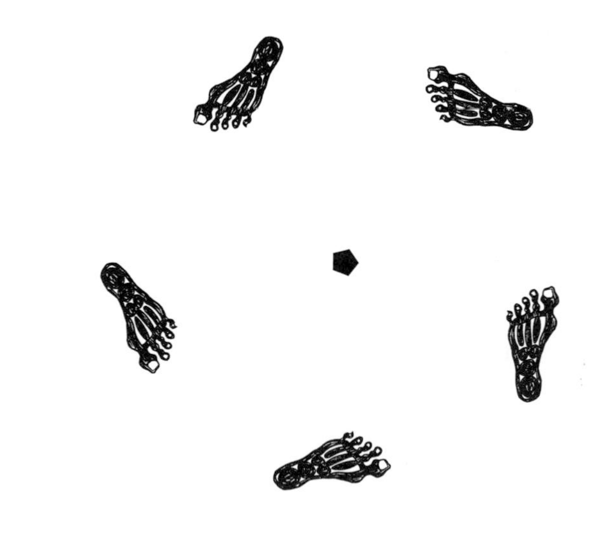

* The operation of rotation by $v\left(\frac{2\pi}{5}\right)$ may be written 5^v.

The multiplication table for the group is

E	5	5^2	5^3	5^4
5	5^2	5^3	5^4	E
5^2	5^3	5^4	E	5
5^3	5^4	E	5	5^2
5^4	E	5	5^2	5^3

As another example consider the effect of adding a mirror line to the previous group. It may be introduced anywhere in the plane as long as it passes through the 5-fold axis—the invariant point. The 5-fold axis acting on the mirror line generates a set of five mirror lines as indicated below.

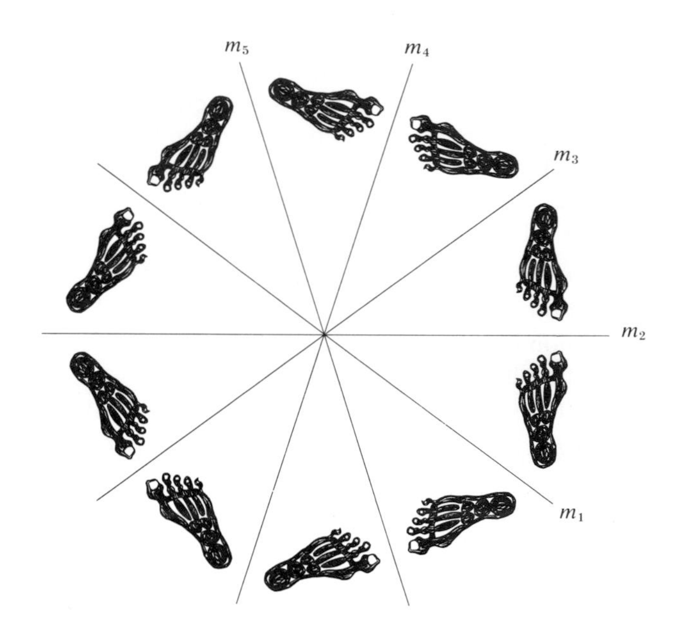

The point group now has ten operations, five rotations and five reflections. This new group $5m$ is of order 10 and has the multiplication table that follows. Several interesting properties of the table are worth study. Although we will not present a multiplication table for each group we consider in this book, it will be instructive to the student to derive several such tables and relate them to the pictorial representations of the groups.

In the present group $5m$, we see that the group operations do not commute; that is, the result of two sequential operations depends on the order of their performance. The operation 5 followed by the operation m_1 is equivalent to the operation m_5:

$$m_1 5 \equiv m_5 \; *$$

*It is common in scientific notation to use $\equiv$ to mean "is identically equal to."

But m_1 followed by 5 is equivalent to m_2:

$$5m_1 \equiv m_2$$

It is also interesting to note that the set of operations including the identity and the rotations forms an independent group of its own. The part of the multiplication table for $5m$ outlined by dotted lines is identical to the multiplication table for 5. The set of mirror planes does not produce an independent group: the product of two mirror planes always produces a rotation. Rotations produce objects of the same handedness. A mirror plane produces an object of the opposite handedness, and consequently in any group two mirror planes must always be equivalent to a rotation.

Second Operation

First Operation

E	5	5^2	5^3	5^4	m_1	m_2	m_3	m_4	m_5
5	5^2	5^3	5^4	E	m_5	m_1	m_2	m_3	m_4
5^2	5^3	5^4	E	5	m_4	m_5	m_1	m_2	m_3
5^3	5^4	E	5	5^2	m_3	m_4	m_5	m_1	m_2
5^4	E	5	5^2	5^3	m_2	m_3	m_4	m_5	m_1
m_1	m_2	m_3	m_4	m_5	E	5	5^2	5^3	5^4
m_2	m_3	m_4	m_5	m_1	5^4	E	5	5^2	5^3
m_3	m_4	m_5	m_1	m_2	5^3	5^4	E	5	5^2
m_4	m_5	m_1	m_2	m_3	5^2	5^3	5^4	E	5
m_5	m_1	m_2	m_3	m_4	5	5^2	5^3	5^4	E

III

CRYSTALLOGRAPHIC SYMMETRY IN TWO DIMENSIONS

The Crystal Lattice

A two-dimensional crystal lattice is a set of equivalent points defined in terms of unit translation vectors **a** and **b** such that if there is a point (x, y) in the set of points, there are also points

$$(x', y') = (x, y) + (n_x\mathbf{a}, n_y\mathbf{b})$$

where n_x and n_y range over all integral values.

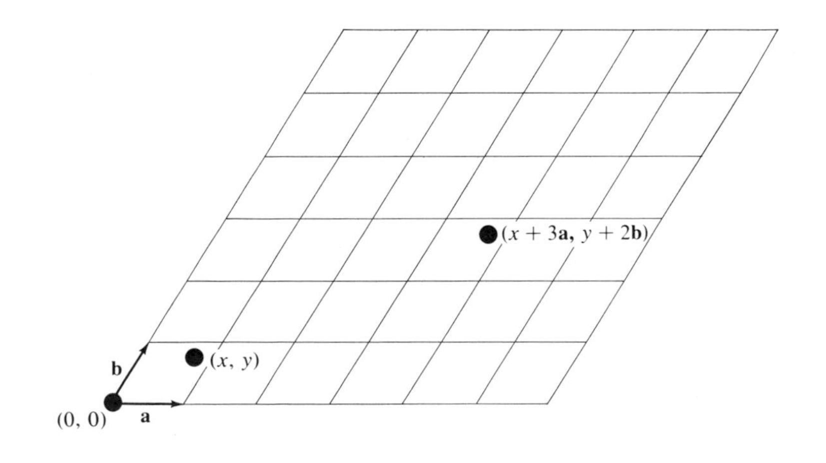

If any one lattice point is chosen as the fixed point of a point group, only certain possible point groups are compatible with the lattice—namely only those that include the following symmetry elements:

$$m \quad 2 \quad 3 \quad 4 \quad 6$$

(As shown on page 35, groups with 5-fold symmetry are not compatible.) Descriptions of these point groups follow. A thorough study of their properties will be helpful in understanding the three-dimensional point groups that are useful in the study of molecular structure and that are discussed in Section IV.

The Ten Crystallographic Plane Point Groups

1
2
m
$2mm$
3
$3m$
4
$4mm$
6
$6mm$

Plane Point Group

1

A single left-handed object. The corresponding point group with a right foot may also exist. The pair are said to be enantiomorphous. The single element of the point group is the identity operation.

Symmetry operations:

$$E: \quad x, y \rightarrow x, y$$

Multiplication table:

$$E$$

Order of group: 1

Plane Point Group

2

Two left feet related by a 2-fold rotation axis.

Symmetry operations:

$$E: \quad x, y \rightarrow x, y$$

$$2: \quad x, y \rightarrow -x, -y$$

Multiplication table:

E	2
2	E

Order of group: 2

Plane Point Group

m

One left foot and one right foot, related by a mirror line.

Symmetry operations:

$$E: \quad x, y \rightarrow x, y$$

$$m: \quad x, y \rightarrow -x, y$$

Multiplication table:

E	m
m	E

Order of group: 2

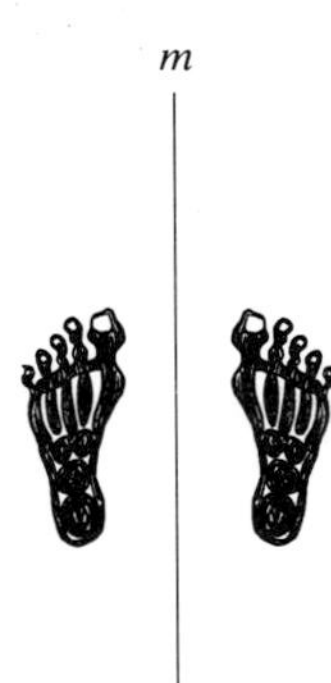

A 2-fold axis combined with a mirror line produces a second mirror line perpendicular to the first. Two left feet and two right feet result. Two elements—the 2-fold axis and one of the mirrors—serve completely to define the group. These two elements are called the *generators* of the group. The other operations follow from the fact that the product of any two group operations is also a group operation. The symbol $2m$ means that there is a mirror line passing through the 2-fold axis. The second m indicates that a second mirror line is perpendicular to the first.

Plane Point Group

2mm

Symmetry operations:

$$E: \quad x, y \rightarrow x, y$$
$$2: \quad x, y \rightarrow -x, -y$$
$$m_x: \quad x, y \rightarrow x, -y$$
$$m_y: \quad x, y \rightarrow -x, y$$

Multiplication table:

E	2	m_x	m_y
2	E	m_y	m_x
m_x	m_y	E	2
m_y	m_x	2	E

Order of group: 4

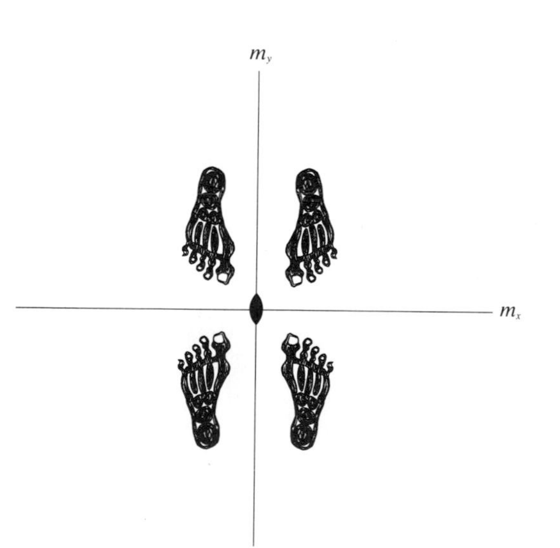

Plane Point Group

3

Three left feet, related by a 3-fold rotation axis. That the multiplication table for this group is symmetric, as it is for many others, is a consequence of the commutative property of the group. The product of any two symmetry operations may be taken in either order to obtain the same result.

Symmetry operations:

$$E:\quad x, y \longrightarrow x, y$$

$$3:\quad x, y \longrightarrow -\frac{1}{2}x - \frac{\sqrt{3}}{2}y,\quad +\frac{\sqrt{3}}{2}x - \frac{1}{2}y$$

$$3^2:\quad x, y \longrightarrow -\frac{1}{2}x + \frac{\sqrt{3}}{2}y,\quad -\frac{\sqrt{3}}{2}x - \frac{1}{2}y$$

Multiplication table:

E	3	3^2
3	3^2	E
3^2	E	3

Order of group: 3

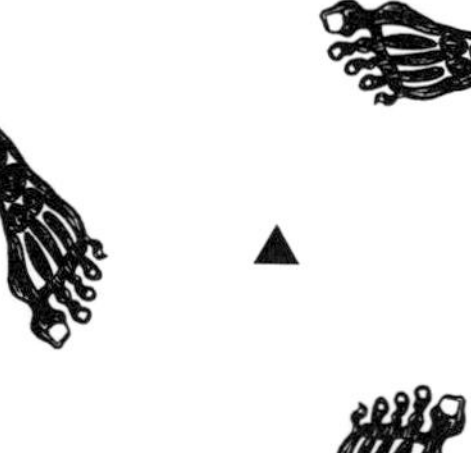

Addition of a mirror plane to group 3 produces three equivalent mirror planes for a total of six symmetry elements. Three right and three left feet are generated.

Plane Point Group

3m

Symmetry operations:

$$E: \quad x, y \rightarrow x, y$$

$$3: \quad x, y \rightarrow x\cos\left(\frac{2\pi}{3}\right) - y\sin\left(\frac{2\pi}{3}\right), x\sin\left(\frac{2\pi}{3}\right) + y\cos\left(\frac{2\pi}{3}\right)$$

$$3^2: \quad x, y \rightarrow x\cos\left(\frac{4\pi}{3}\right) - y\sin\left(\frac{4\pi}{3}\right), x\sin\left(\frac{4\pi}{3}\right) + y\cos\left(\frac{4\pi}{3}\right)$$

$$m_1: \quad x, y \rightarrow -x\cos\left(\frac{4\pi}{3}\right) + y\sin\left(\frac{4\pi}{3}\right), x\sin\left(\frac{4\pi}{3}\right) + y\cos\left(\frac{4\pi}{3}\right)$$

$$m_2: \quad x, y \rightarrow -x\cos\left(\frac{2\pi}{3}\right) + y\sin\left(\frac{2\pi}{3}\right), x\sin\left(\frac{2\pi}{3}\right) + y\cos\left(\frac{2\pi}{3}\right)$$

$$m_3: \quad x, y \rightarrow -x, y$$

Multiplication table:

E	3	3^2	m_1	m_2	m_3
3	3^2	E	m_3	m_1	m_2
3^2	E	3	m_2	m_3	m_1
m_1	m_2	m_3	E	3	3^2
m_2	m_3	m_1	3^2	E	3
m_3	m_1	m_2	3	3^2	E

Order of group: 6

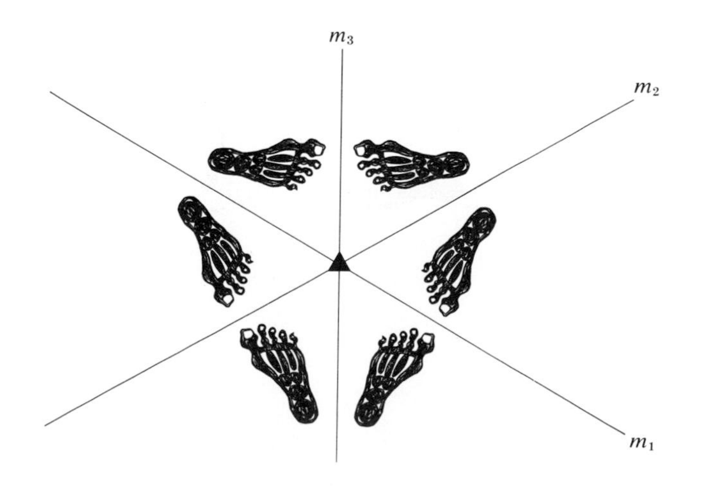

Plain Point Group

4

Four left feet, related by a 4-fold rotation axis. The operation 4^2 is equivalent to 2.

Symmetry operations:

$$E: \quad x, y \rightarrow x, y$$
$$4: \quad x, y \rightarrow -y, x$$
$$4^2: \quad x, y \rightarrow -x, -y$$
$$4^3: \quad x, y \rightarrow y, -x$$

Multiplication table:

E	4	4^2	4^3
4	4^2	4^3	E
4^2	4^3	E	4
4^3	E	4	4^2

Order of group: 4

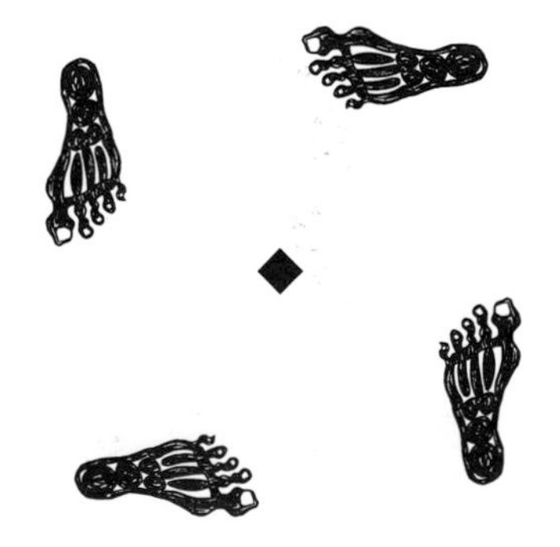

Plane Point Group

4mm

Addition of a mirror line to group 4 produces a pattern of four left feet and four right feet marching in opposite directions. Mirror planes are generated halfway between the x and y axes, as indicated by the second m in the group symbol.

Symmetry operations:

E:	$x, y \rightarrow x, y$
4:	$x, y \rightarrow -y, x$
4^2:	$x, y \rightarrow -x, -y$
4^3:	$x, y \rightarrow y, -x$
m_x:	$x, y \rightarrow x, -y$
m_y:	$x, y \rightarrow -x, y$
m_1:	$x, y \rightarrow y, x$
m_2:	$x, y \rightarrow -y, -x$

Multiplication table:

E	4	4^2	4^3	m_x	m_y	m_1	m_2
4	4^2	4^3	E	m_2	m_1	m_x	m_y
4^2	4^3	E	4	m_y	m_x	m_2	m_1
4^3	E	4	4^2	m_1	m_2	m_y	m_x
m_x	m_1	m_y	m_2	E	4^2	4	4^3
m_y	m_2	m_x	m_1	4^2	E	4^3	4
m_1	m_y	m_2	m_x	4^3	4	E	4^2
m_2	m_x	m_1	m_y	4	4^3	4^2	E

Order of group: 8

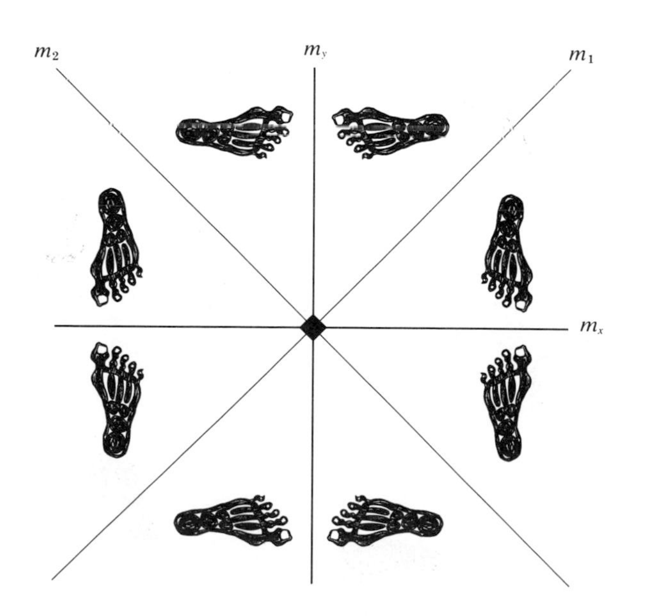

Plane Point Group

6

Six left feet, related by a 6-fold rotation axis.

Symmetry operations: $E \quad 6 \quad 6^2 \quad 6^3 \quad 6^4 \quad 6^5$

The student should now be able to write the exact expressions for the coordinate transformations associated with each symmetry element. Note that

$$6^2 \equiv 3$$

$$6^4 \equiv 3^2$$

$$6^3 \equiv 2$$

so that we could and often do write the symmetry elements as

$$E \quad 6 \quad 3 \quad 2 \quad 3^2 \quad 6^5$$

Multiplication table:

E	6	6^2	6^3	6^4	6^5
6	6^2	6^3	6^4	6^5	E
6^2	6^3	6^4	6^5	E	6
6^3	6^4	6^5	E	6	6^2
6^4	6^5	E	6	6^2	6^3
6^5	E	6	6^2	6^3	6^4

Order of group: 6

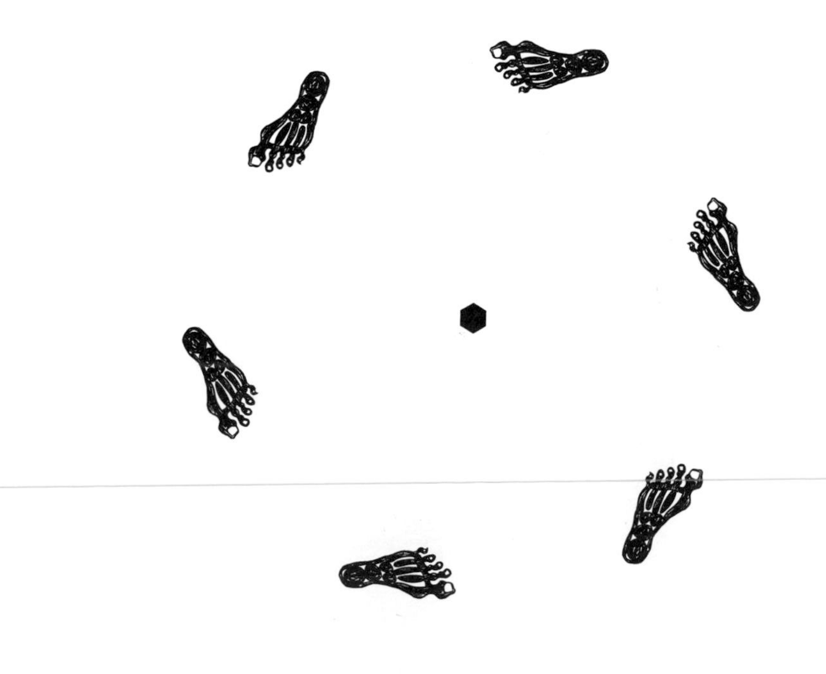

Plane Point Group

6mm

Addition of a mirror plane to group 6 produces six mirror planes at angles of 30° with one another. Six left feet and six right feet are produced.

Symmetry operations:

$$E \quad 6 \quad 6^2 \quad 6^3 \quad 6^4 \quad 6^5 \quad m_1 \; m_2 \; m_3 \; m_4 \; m_5 \; m_6$$

The student should now be able to write the multiplication table, as well as the exact expressions for the coordinate transformations associated with each symmetry operation.

Order of group: 12

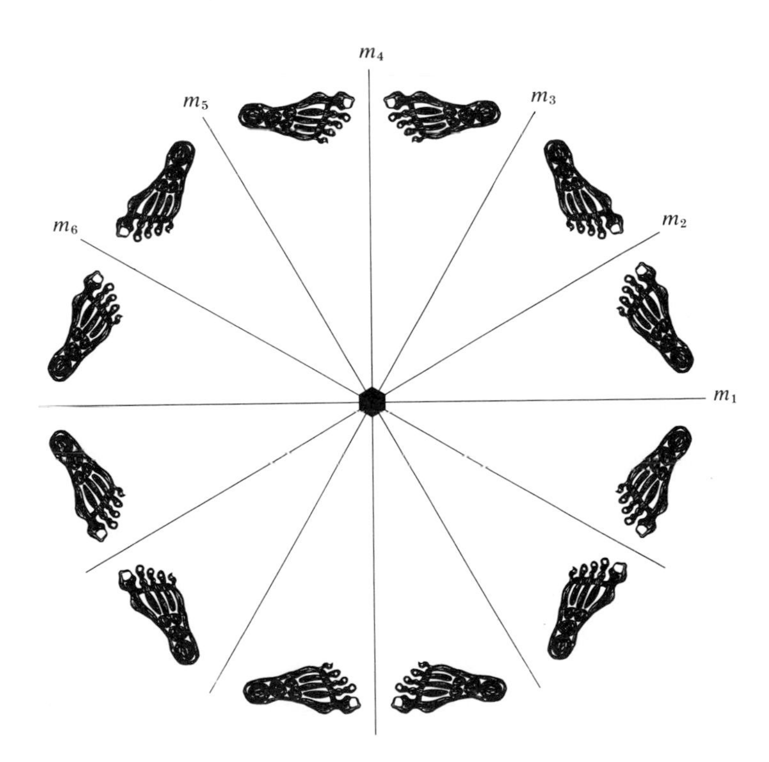

Crystallographic Plane Groups

A crystal of any dimensionality is defined by a group of translational symmetry operations. In two dimensions we have

$$x' = x + n_x$$

$$y' = y + n_y$$

where n_x and n_y can assume any integer values; the order of the group is infinity, and the identity element is that for which $n_x = n_y = 0$. (For translational operations, we express x and y in units of the lengths of **a** and **b**, the unit cell translation vectors, as we did on page 41). The translation group so defined is not a point group, since the general translation operation leaves no point invariant.

We may, however, define a group that contains as its elements the elements of a translation group *and* those of a point group. Such groups are known in two dimensions as crystallographic *plane groups* and in three dimensions as crystallographic *space groups*. In these groups the crystal lattice has the symmetry of the generating point group but the entire contents of the crystal need not. There are often sets of points invariant under one or more operations of the point group, but the symmetry about these points is not necessarily the same. (Problems 26 and 27 on page 166 illustrate this.)

Aside from the integral translations that define the crystal lattice, there can exist in plane and space groups a new class of symmetry operations, which involve fractional translations such as

$$x' = x + \frac{1}{2}$$

An example is discussed on page 61. Nonintegral translation operations are of great importance in three-dimensional crystals. (See problem 33, page 169.)

In two dimensions, there are only seventeen possible crystallographic plane groups. We present a few of these in the following pages.

A Few of the Crystallographic Plane Groups

$P1$
$P2$
Pm
Pg

Crystallographic Plane Group

$P1$

The point group symmetry is 1. All feet are left-handed. Directions and lengths of the unit translations are not restricted.

Three different identical repeating units (unit cells) are shown to demonstrate that the origin is arbitrary. It is required only that each unit contain one complete foot (perhaps made up of pieces of separate feet) and that the surroundings of each cell corner—lattice point—be identical.

The symbol P indicates that the lattice is primitive: there is only one lattice point per unit cell. We have chosen to use primitive lattices in our examples, but to emphasize the symmetry of a translation group it is often convenient to choose a larger unit cell than the smallest repeating unit.

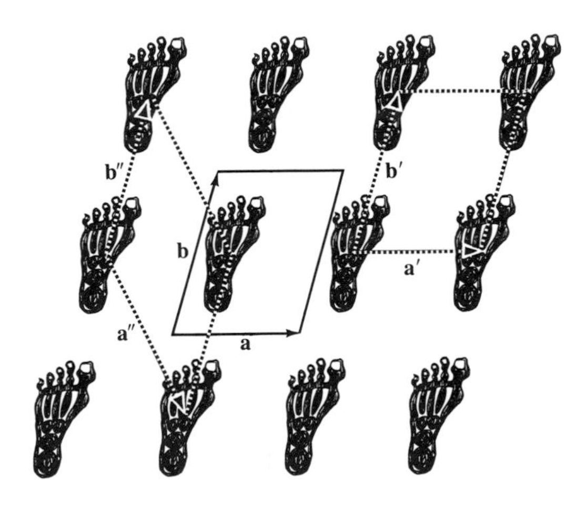

Crystallographic Plane Group

P2

The point group is 2. A 2-fold axis lies between every pair of feet located at lattice points and halfway between lattice points. Feet of one chirality only are present. There are no restrictions on the directions and lengths of the unit translations.

b

a

Crystallographic Plane Group

The point group is m. Two nonequivalent mirror planes are perpendicular to **a**, one at the half-translational point. The two unit translation axes are required to be perpendicular to each other, but there is no restriction on their lengths. Half of the feet are right and half are left.

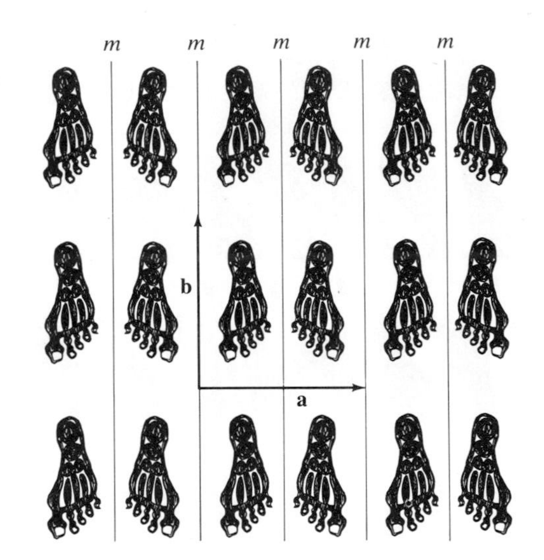

Crystallographic Plane Group

Pg

A type of symmetry element that exists only in translation groups is illustrated with this group, namely the operation defined by

$$g: \quad x, y \rightarrow x + \frac{1}{2}, \frac{1}{2} - y$$

This is equivalent to a mirror line at $y = \frac{1}{4}$ followed by a translation of one-half the unit translation along x.

The operation is known as a glide and, like the mirror, it produces a left foot from a right foot. Glide operations are very important in the three-dimensional crystallographic space groups. Although we will not discuss these in detail, the figure appearing on the next page illustrates the geometry of such an operation in three dimensions. Combination of all possible three-dimensional point groups and translational symmetries including glide operations produces 230 space groups.

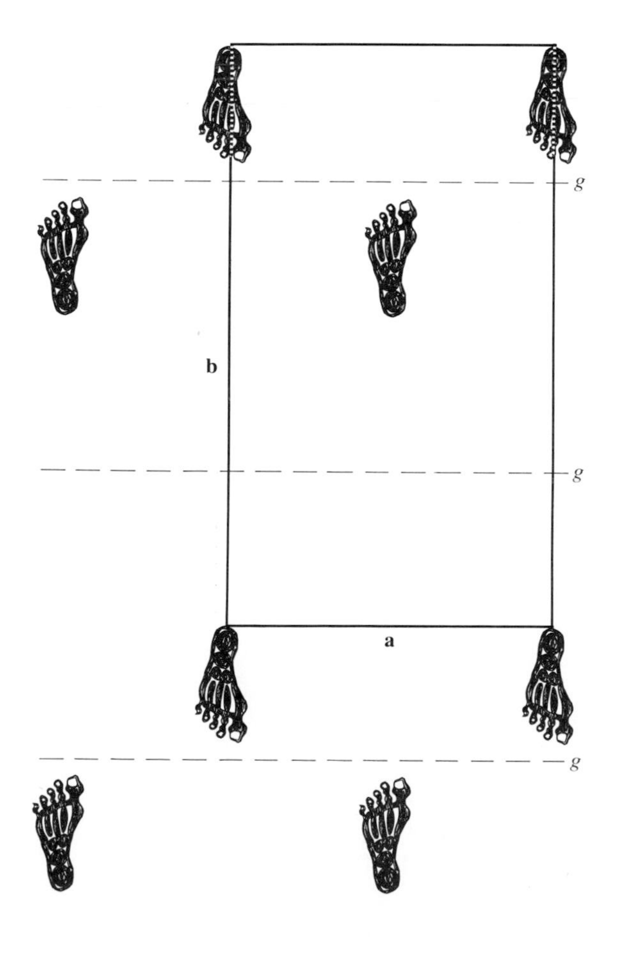

g = glide planes

x and y can be measured along translational vectors **a** and **b**, respectively

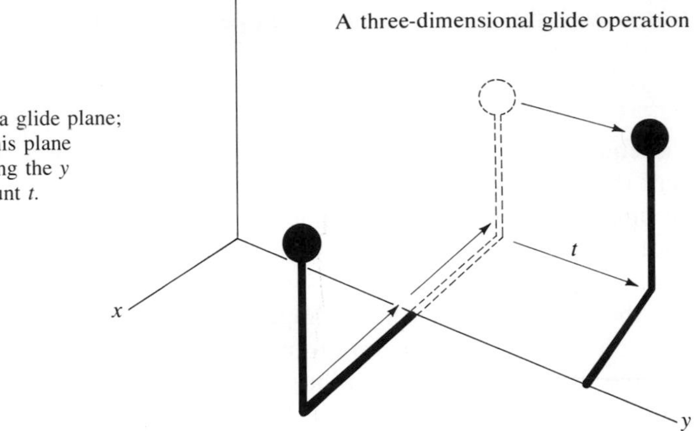

A three-dimensional glide operation

The yz plane is a glide plane; a reflection in this plane is translated along the y axis by an amount t.

IV

THE THREE-DIMENSIONAL POINT GROUPS

Symmetry Groups in Three Dimensions

Symmetry operations in three dimensions generate the three-dimensional point groups. We have introduced two of these operations—the mirror plane and the rotation axis—in our discussion of the two-dimensional point groups. In the third dimension, we must introduce a new operation—the rotation-reflection axis—which is defined on page 93. Only thirty-two unique point groups are compatible with the translational symmetry of a crystal lattice; these are the crystallographic point groups. In this section we will describe these point groups along with a number of other noncrystallographic point groups that sometimes occur in descriptions of molecular structure.

Various notations have been used to describe point groups. Detailed explanations of the notations will appear in connection with the presentation preceding the individual point groups. It behooves the student to become familiar with these notations as an aid in reading the chemical, crystallographic, and molecular structure literature. The notation most commonly adopted by chemists is the Schoenflies notation, in which the name of a group is usually derived from its principal rotation axis. Accordingly, the groups C_n have only a rotation axis of order n. Mirror planes perpendicular to the principal axis (*horizontal* mirror planes) are denoted by appending a subscript h, for example, C_{2h}. Mirror planes that include the rotation axis (*vertical* mirror planes) are denoted by an appended subscript v, for example, C_{3v}. Groups with 2-fold axes perpendicular to the principal axis are denoted by D (for *dihedral*), again possibly with subscripts v and h. A subscript d indicates a mirror plane that bisects the angle between two 2-fold axes and includes the principal axis. The designation S is given to groups that include as their principal symmetry element a rotation-reflection axis. Special designations containing O and T are given to the cubic point groups. The uses of these symbols will become clear on examination of the individual point groups.

An alternative notation preferred by crystallographers is the Hermann-Mauguin notation, which depends less on special symbols such as C, D, S, T, and O, but is more systematic in giving the information necessary to generate the symmetry operations of the point group. Accordingly, the first part of the symbol is simply the number that gives the order of the principal rotation axis. Following symbols denote the presence of mirror planes or subsidiary rotation axes.

In our presentation of the point groups, we give a brief description of the symmetry elements of the group and illustrate these symmetry elements by a three-dimensional stereoscopic drawing of a completely asymmetric object, operated on by the symmetry elements of the group. We then present a description and a picture of a molecule (sometimes hypothetical, as we explained on page 3) that has the symmetry of the point group.

Point Groups with Pure Rotation Axes

C_n n

These point groups each consist of only a single family of symmetry operations—those generated by a single rotation axis. If the rotation axis is of order n, the Schoenflies notation is

$$C_n$$

The alternative Hermann-Mauguin notation is simply

$$n$$

Although point groups with any value of n are possible, we include here only the crystallographic point groups plus the noncrystallographic group C_5.

Schoenflies Symbol	Hermann-Mauguin Symbol
C_1	1
C_2	2
C_3	3
C_4	4
C_5	5
C_6	6

C_1 1

There is no symmetry operation that can bring this asymmetric figure into coincidence with itself. The identity operation is the only operation in the point group. It is convenient to classify this group as a pure rotation group where the only operation is a rotation by 2π about an arbitrary axis. A molecule with this symmetry would be **optically active**, and both right- and left-handed forms would exist.

Order of group: 1

Crystallographic symmetry: Triclinic

Xenon(II) fluoride fluorosulfate
$FXeSO_3F$

Before the preparation of the compounds described in this paper no compounds of Xe(II) were known with **ligands** other than fluoride. The compound illustrated here contains an —O—Xe—F fragment and thus constitutes the first structurally well-characterized example of a Xe(II) compound with oxygen bonds.

N. Bartlett, M. Wechsberg, F. O. Sladsky, P. S. Bulliner, G. R. Jones, and R. D. Burbank, *Chemical Communications*, 703 (1969).

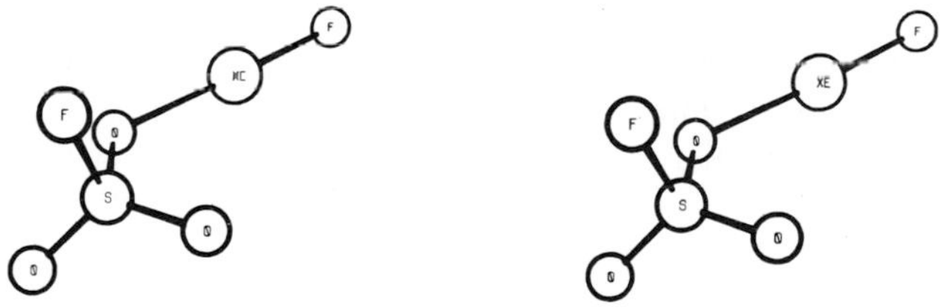

C_2 2

Aside from the identity operation, this point group has a single symmetry operation: a 2-fold axis. This is the vertical coordinate axis in the drawing. Both objects are of the same handedness; they may be superimposed simply by a rotation around the 2-fold axis. A molecule with this point group symmetry would be optically active.

Order of group: 2

Crystallographic symmetry: Monoclinic

2,2′-Biphenyldisulfide
$C_{12}H_8S_2$

This substance has been found capable of readily losing the disulfide (S_2) fragment in a variety of chemical reactions. Thus, it was desirable to study it to determine whether or not the C—S bonds were unusually long, that is, weak. Our results do not indicate that such a simple correlation can be made between the chemical behavior and the geometrical structure.

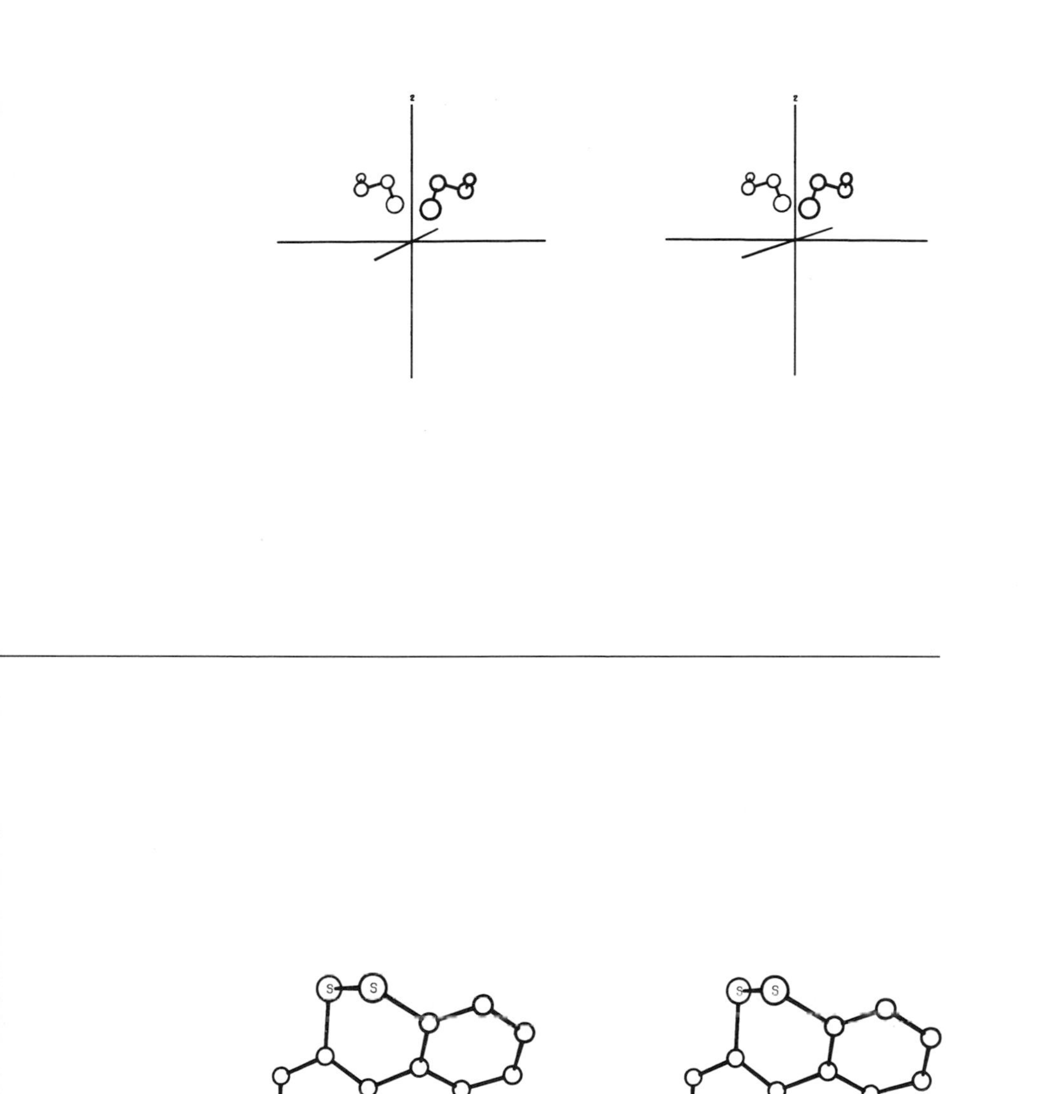
S
S
S
S

C_3 3

The three symmetry operations in the point group are generated by the 3-fold axis that is perpendicular to the plane of the page. The symmetry operations are a rotation by $2\pi/3$ or 120°, a rotation by $2(2\pi/3)$ or 240°, and a rotation by $3(3\pi/3)$ or 360°, which is, of course, the identity operation. All of the objects are of the same handedness, and a molecule with this point group symmetry would therefore be optically active.

Order of group: 3

Crystallographic symmetry: Trigonal

Tris (2-dimethylaminoethyl)aminecobalt(II) bromide cation $\left[\left((CH_3)_2NCH_2CH_2\right)_3N\right]CoBr^+$

Five bonds are formed to cobalt in this **complex cation**—four from the **amine** and one from a bromide **ion**. As shown in the drawing the amine ligand looks like an umbrella: three $—CH_2—CH_2—N(CH_3)_2$ fragments radiate out from a common nitrogen atom. Although a number of **five-coordinated** cobalt(II) complexes have been described, what the factors are that determine the stereochemistry of these compounds is not well understood. In the compound discussed here, the 3-fold symmetry around the metal atom is probably determined by the constraints imposed by the geometry of the ligand molecule. The hydrogen atoms are not labeled in this and most of the following drawings.

M. Di Vaira and P. L. Orioli, *Inorganic Chemistry*, **6**, 955 (1967).

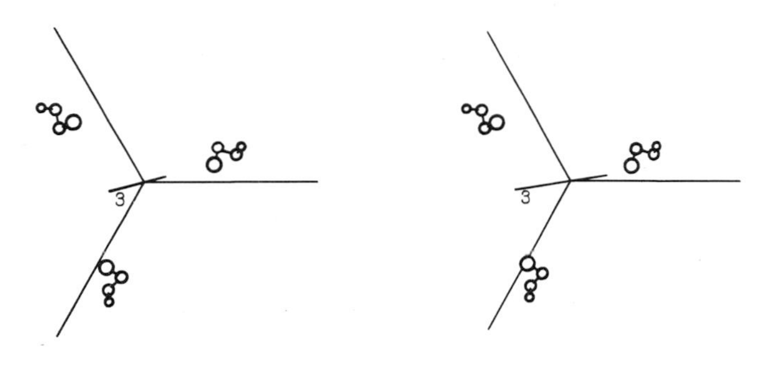
3
3

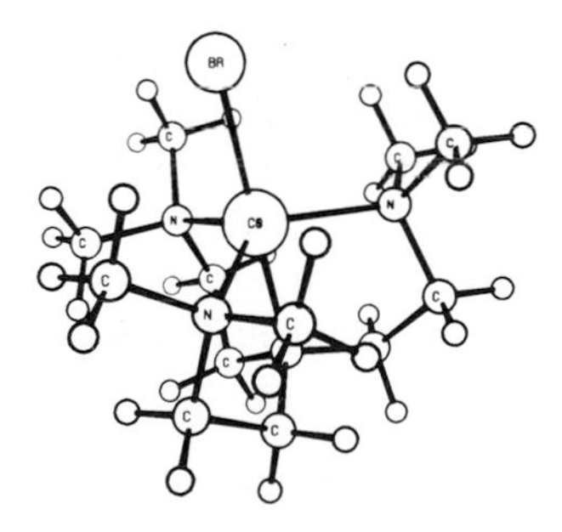
BR

BR

C_4 4

A 4-fold rotation axis generates the four operations of this point group: a rotation by $2\pi/4$, a rotation by $2(2\pi/4)$, a rotation by $3(2\pi/4)$ and, finally, a rotation of $4(2\pi/4)$, which is the identity operation. All of the objects are of the same handedness and a molecule with this symmetry would be optically active.

Order of group: 4

Crystallographic symmetry: Tetragonal

Tetramethylcyclobutadiene
$(CH_3)_4C_4$

In a possible structure for this unstable molecule the four carbon-carbon bond lengths in the four-membered ring are equal in length. None of the hydrogen atoms of the methyl group (CH_3) are in the plane of the ring. **Quantum** mechanical calculations suggest that when free, the molecule does not have even approximate 4-fold symmetry, although it and similar molecules are usually found to be symmetrical when bonded to metal atoms.

R. P. Dodge and V. Schomaker, *Acta Crystallographica*, **18**, 614(1965).

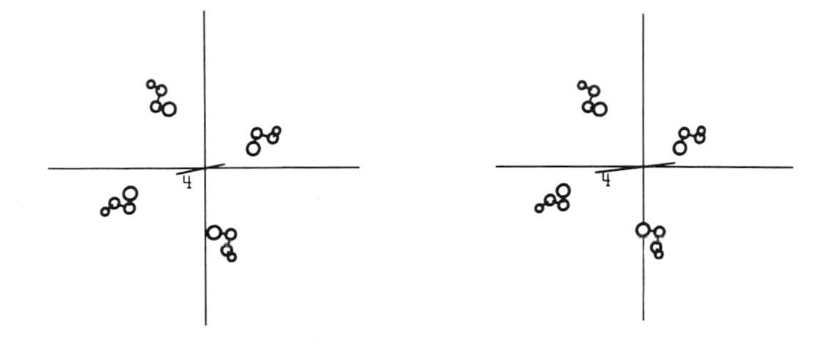
4
4

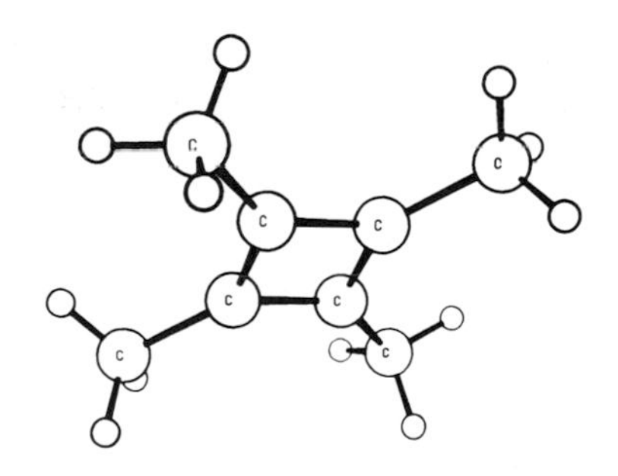

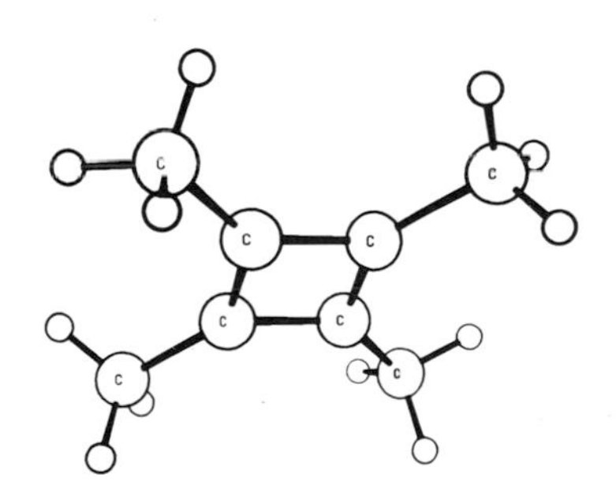

C_5 5

The principal symmetry element is a 5-fold axis, and the five symmetry operations are subsequent rotations of $2\pi/5$, the last of which is, of course, the identity operation. (The five lines shown perpendicular to the 5-fold axis in the figure are not symmetry axes.) All of the objects are of the same handedness; a molecule with this symmetry would be optically active. A crystal structure cannot belong to point group C_5, nor can it belong to any of the point groups based upon C_5.

Order of group: 5

Crystallographic symmetry: none

Pentamethyl cyclopentadienyl radical
$(CH_3)_5C_5$

As with the parent cyclopentadienyl (C_5H_5) **radical** (see page 122), this unstable species may form a **delocalized** bond along the 5-fold axis to a metal ion.

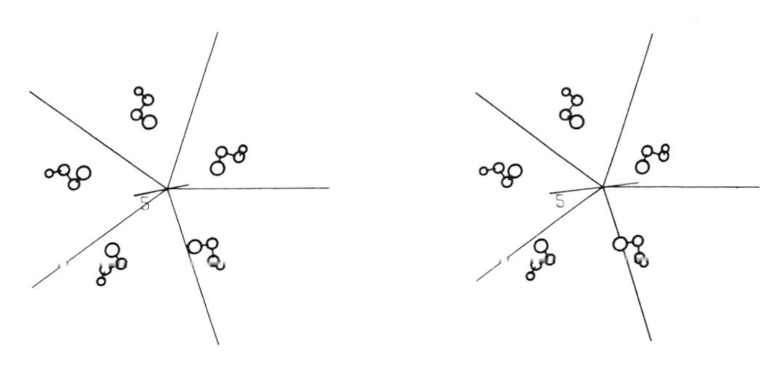
5
5

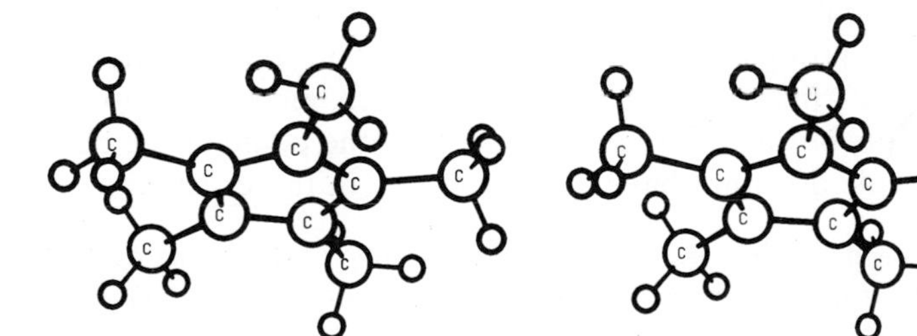
C

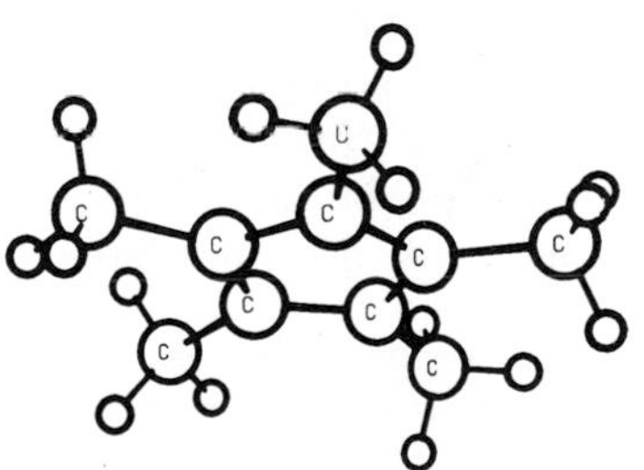

C_6 6

The generating symmetry element is a 6-fold rotation axis. The six operations of the group are the subsequent rotations by $2\pi/6$. All of the objects are of the same handedness; a molecule with this symmetry would be optically active.

Order of group: 6

Crystallographic symmetry: Hexagonal

Hexamethylbenzene
$(CH_3)_6C_6$

The symmetry of this molecule depends upon the angles of rotation of the methyl groups about the carbon-carbon bonds linking them to the benzene ring. In this point group, each methyl group has an identical rotation, but no hydrogen atoms are in the plane of the benzene ring. The molecule does not have this symmetry in the crystal.

K. Lonsdale, *Proceedings of the Royal Society* (London) **A123**, 494 (1929).

L. O. Brockway and J. M. Robertson, *Journal of the Chemical Society* (London), 1324 (1939).

W. C. Hamilton, J. Edmonds, A. Tippe, and J. J. Rush, *Discussions of the Faraday Society*, **48**, 192 (1969).

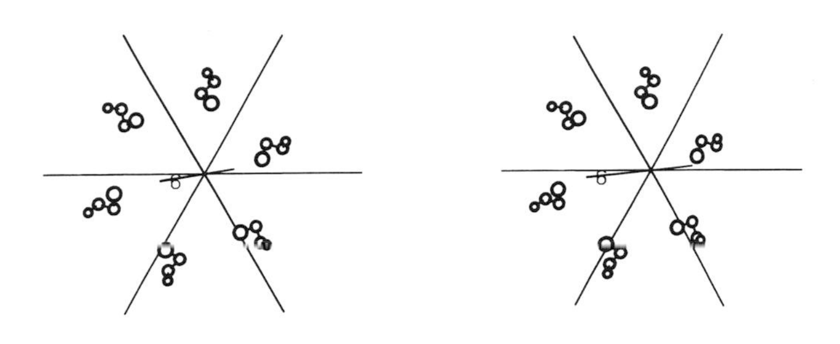

Point Groups with a Single Rotation Axis That Lies in a Mirror Plane

C_{nv} nm

In these point groups, a mirror plane is parallel to—and includes—the principal symmetry axis. Since the principal symmetry axis is usually considered to be in a vertical direction, the mirror plane is called a *vertical* mirror plane and is denoted by a subscript v. The Schoenflies symbols are thus C_{nv}. The alternative Hermann-Mauguin notation first gives the principal axis, followed by an m to indicate the mirror plane. The n-fold axis generates n such mirror planes. If n is even, the rotation by π creates an identical mirror plane, but in these cases, there is always a second set of mirror planes halfway between those related by the n-fold axis. Thus there are always n distinct mirror planes in the groups C_{nv}. The point group symbol for odd n is

$$nm$$

and for even n is

$$nmm$$

the second m denoting the set of mirror planes that bisect the principal set.

Schoenflies Symbol	Hermann-Mauguin Symbol
C_{2v}	$2mm$*
C_{3v}	$3m$
C_{4v}	$4mm$
C_{5v}	$5m$
C_{6v}	$6mm$

*Often called $mm2$ by crystallographers.

C_{2v} 2mm

Two mirror planes perpendicular to one another are contained in this point group; one is defined by the x and z axes; the other is defined by the y and z axes. The line of intersection of the two mirror planes generates the 2-fold axis that lies along z. This direction is usually taken as the vertical axis. Thus there are four symmetry elements in the point group: the identity element, the 2-fold axis, and the two mirror planes. The mirror planes generate both right- and left-handed objects; a molecule having this symmetry would not be optically active.

Order of group: 4

Crystallographic symmetry: Orthorhombic

Bis(trimethylamine)oxovanadium dichloride $[(CH_3)_3N]_2(VO)Cl_2$

The 2-fold axis lies along the V—O bond and relates the two chlorides (Cl) and also the two trimethylamines $[(CH_3)_3N]$ to each other. The hydrogen atoms of the methyl groups are not shown. The precise geometry of **bis**(trimethylamine)oxovanadium dichloride was studied in order to aid in the interpretation of optical spectra for low symmetry vanadium compounds.

J. E. Drake, J. Vekris, J. S. Wood, *Journal of the Chemical Society* (London), [A] 1000 (1968).

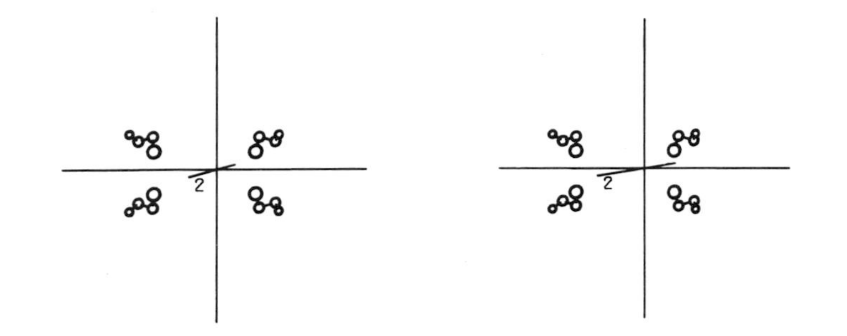
2
2

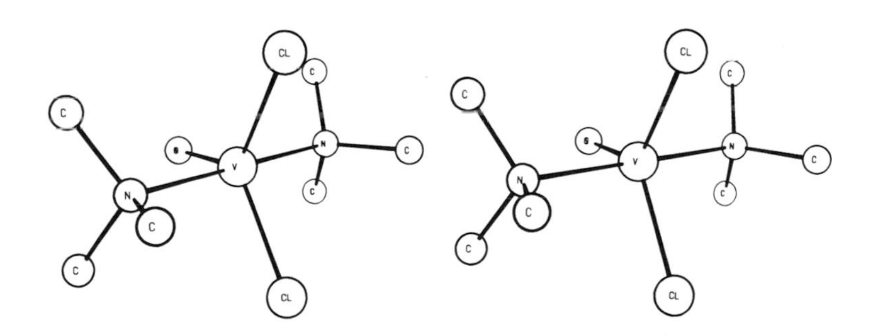
CL
C
C
N
V
N
C
C
C
C
CL
CL
C
C
N
V
N
C
C
C
C
CL

C_{3v} 3m

This point group is produced by the combination of a 3-fold axis and three mirror planes; each mirror plane includes the 3-fold axis and makes angles of 120° with the others. The operation of rotation about the 3-fold axis followed by reflection in a mirror plane is equivalent to reflection in another mirror plane. Objects of both right and left handedness appear, and as in all such enantiomorphic pairs, a molecule with this symmetry could not be optically active.

Order of group: 6

Crystallographic symmetry: Trigonal

Chloroform
CCl_3H

The three-fold axis lies along the C—H bond, and the three mirror planes each contain an H—C—Cl fragment.

S. N. Ghosh, R. Trambarulo, and W. Gordy, *Journal of Chemical Physics*, **20**, 605 (1952).

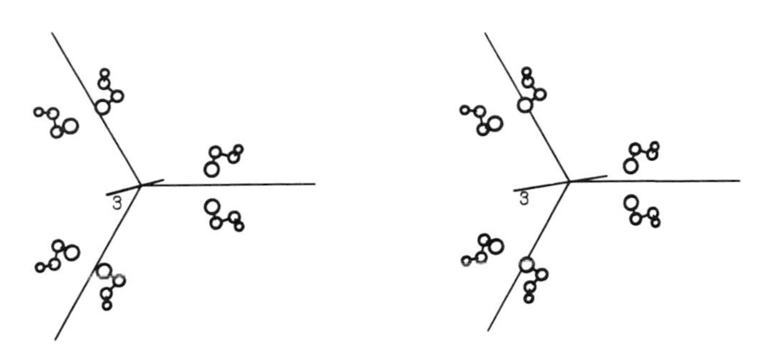
3
3

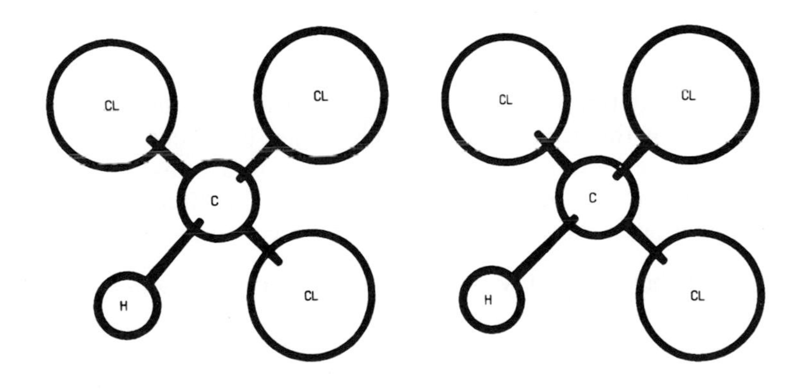
CL
CL
C
H
CL
CL
CL
C
H
CL

C_{4v} 4mm

Four vertical mirror planes are added to the point group C_4 to obtain this point group. One mirror plane includes the x and z axes, another the y and z axes; in combination with the 4-fold axis these two produce another two mirror planes midway between the first pair. Both right- and left-handed objects exist.

Order of group: 8

Crystallographic symmetry: Tetragonal

Pentaborane
B_5H_9

The elucidation of the structure of the boron hydrides was one of the most exciting accomplishments in structural chemistry. Many models, most of them based on sensible ideas, had been suggested. The x-ray structure determination of this compound showed that the geometry was that of a **tetragonal pyramid** and exhibited the B—H—B **bridge bonds** later shown to be typical of the boron hydrides.

W. J. Dulmage and W. N. Lipscomb, *Acta Crystallographica*, **5**, 260 (1952).

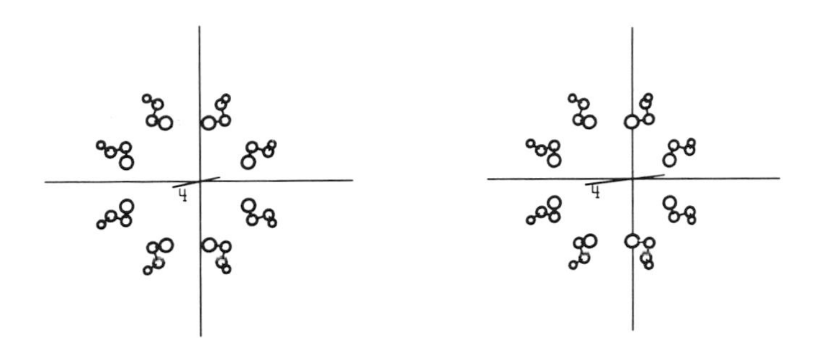

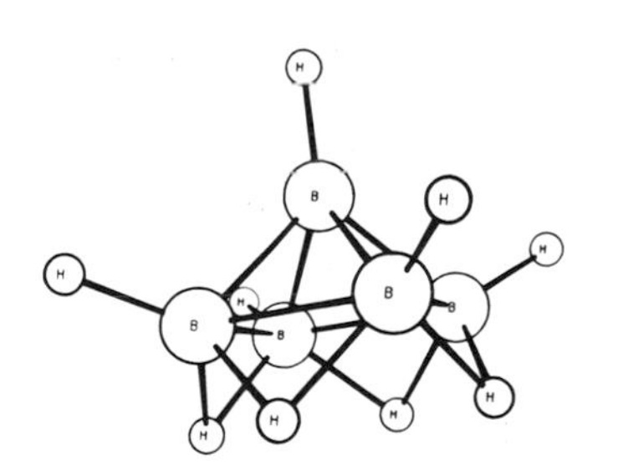

H
B
H
H
B
B
B
H
H
H
H
H

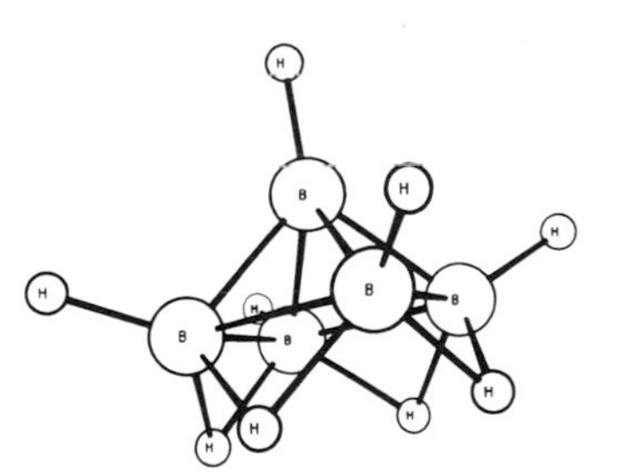

H
B
H
H
B
B
B
H
H
H
H
H

C_{5v} $5m$

This point group is obtained from the point group C_5 by adding mirror planes that include the 5-fold axis; five mirror planes thus result, producing the ten symmetry operations of the point group. Both right- and left-handed molecules exist.

Order of group: 10

Crystallographic symmetry: none

Tridecahydroundecaborate ion
$(B_{11}H_{13})^{-2}$

Fragments of a molecule often have simple symmetries, although the whole molecule may not. Boron compounds often have 5-fold axes, and parts of the structures appear to be fragments of regular **icosahedra**. The ion illustrated here has C_{5v} symmetry if the two bridging hydrogen atoms in the lower part of the figure are ignored.

C. J. Fritchie, Jr., *Inorganic Chemistry*, **6**, 1199 (1967).

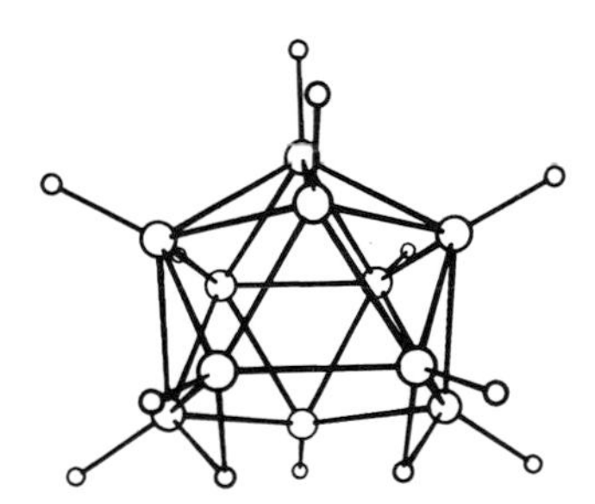

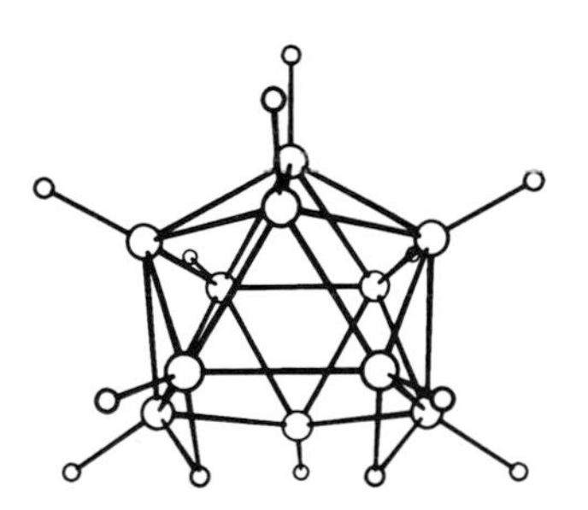

C_{6v} 6mm

Vertical mirror planes are added to the point group C_6. Mirror planes thus occur at intervals of 30° around the principal symmetry axis. There are twelve symmetry operations—six rotations and six reflections. Objects of both chiralities occur.

Order of group: 12

Crystallographic symmetry: Hexagonal

Hexa(chloromethyl)benzene
$C_6(CH_2Cl)_6$

The structure of this molecule has not been determined. The compound consists of six —CH_2Cl fragments bonded to the six carbons of a benzene ring. The structure would be greatly influenced by the large size of the chlorine atoms; the C_{6v} structure shown might be expected to minimize the sum of repulsions between Cl and Cl and Cl and H, although the **steric energy** could be so large that the molecule would become nonplanar and thus lose its 6-fold axis. For further discussion of the effect of multiple chloro substitution on molecular properties see: R. Rudman, *Journal of the Chemical Society* (London) [D], 536 (1970).

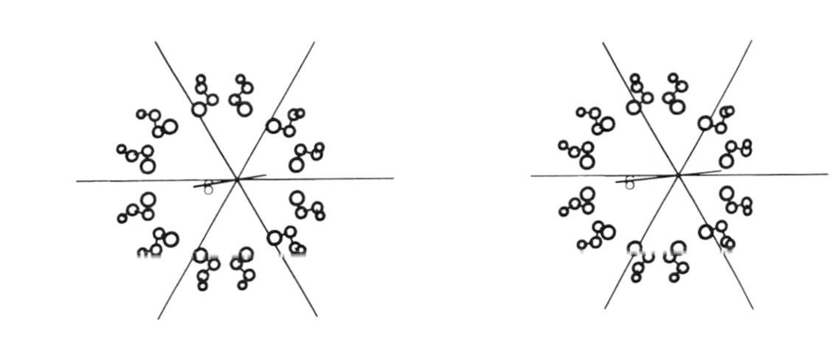
6
6

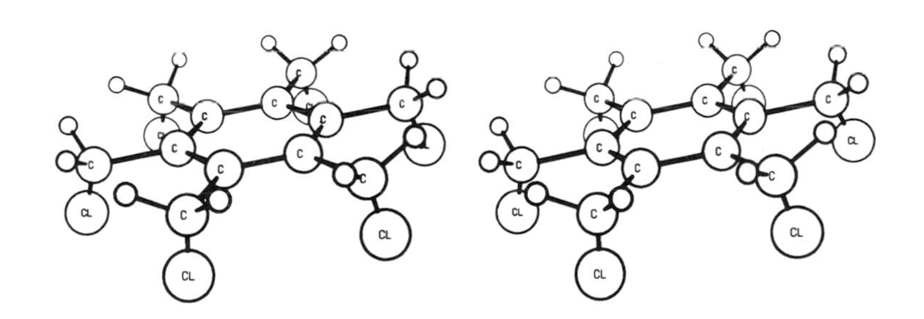
C
CL

Point Groups with Only Rotation-Reflection Axes

S_n $\bar{n}$

In these point groups, the basic symmetry operation is the rotation–reflection axis S_n defined by a rotation of $2\pi/n$ followed by a reflection in a plane perpendicular to the principal symmetry axis. For such a description, the point group has the Schoenflies symbol

$$S_n$$

Instead of choosing the rotation–reflection axis as the third basic symmetry operation necessary to generate three-dimensional point groups, it is also possible to choose the rotation–inversion axis, which is defined as a rotation by $2\pi/n$ followed by inversion through the origin, by which a point x, y, z is transformed to the point $-x$, $-y$, $-z$. The Hermann-Mauguin notation for a rotation–inversion axis is simply

$$\bar{n}$$

The values of n are not necessarily the same in the Schoenflies and Hermann-Mauguin symbols. S_2 is equivalent to $\bar{1}$, and S_6 is identical to $\bar{3}$. Also, S_3 is identical to $\bar{6}$ but is not included in this chapter because it also contains a horizontal mirror plane. The Schoenflies notation C_{ni} is sometimes used to indicate the presence of an n-fold axis plus an inversion center.

Schoenflies Symbol	Hermann-Mauguin Symbol
S_2, C_i	$\bar{1}$
S_4	$\bar{4}$
S_6, C_{3i}	$\bar{3}$

S_2 $\bar{1}$

The single nonidentity symmetry element is the center of symmetry or inversion center that lies in the drawing at the intersection of the three coordinate axes. This is identical to a rotation $2\pi/2$ around an arbitrary axis, followed by reflection in a plane perpendicular to that axis. This group is thus the simplest of those generated by rotation–reflection axes. One of the two objects is right-handed, the other left-handed. As with all other point groups containing S_n axes, a molecule with S_2 symmetry cannot be optically active.

Order of group: 2

Crystallographic symmetry: Triclinic

---

trans-15,16-Diethyldihydropyrene
$C_{20}H_{20}$

Spectroscopic data obtained from this substance suggested that it is **aromatic**, and most aromatic systems are planar. The crystal structure revealed that the compound is strained (nonplanar) but that the bond lengths in the periphery of the molecule are typical of those in aromatic compounds. In order to visualize the inversion center, consider the fact that the ethyl groups (the two carbon chains sticking out from the flat, pancake-like, aromatic pyrene system) are on opposite sides of the pyrene fragment.

A. W. Hanson, *Acta Crystallographica*, **23**, 476 (1967).

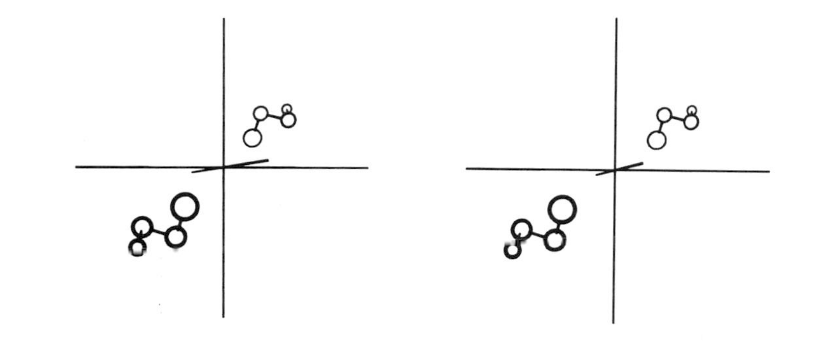

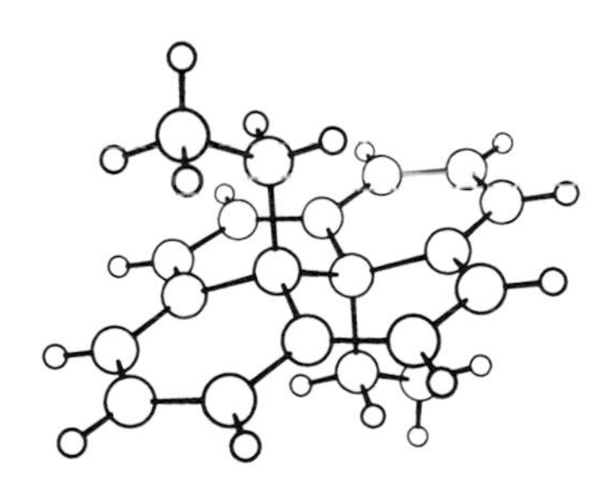

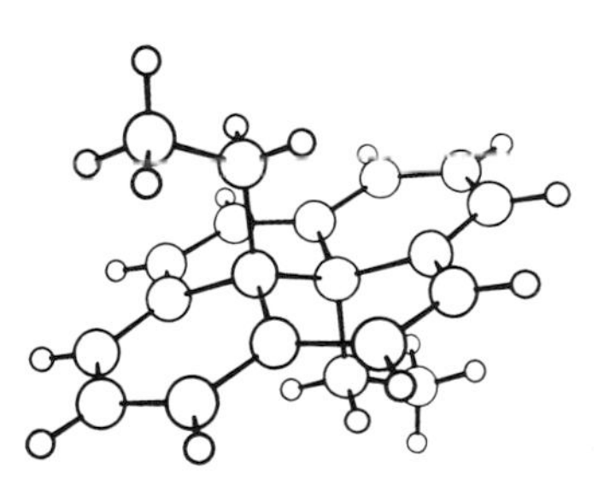

S_4 $\bar{4}$

This point group is generated by a 4-fold rotation–reflection axis. The four operations of the point group may be considered as sequential rotations through 90°, each followed by reflection in the plane perpendicular to the axis. This does not imply that a mirror plane is one of the operations, nor does the notation $\bar{4}$ imply that there is a center of symmetry. Note that two elemental $\bar{4}$ operations applied in turn result simply in a 2-fold rotation. The two objects below the x, y plane are of one handedness, those above are of another.

Order of group: 4

Crystallographic symmetry: Tetragonal

Bis(dipivaloylmethanido)zinc(II)
$Zn\left[(CH_3)_3C—CO—CH—CO—C(CH_3)_3\right]_2$

Hydrogen atoms on the **beta-diketone** ligands are not shown. The tetrahedral configuration around Zn is typical of that produced by a 4-fold rotation–reflection axis. The compound was studied to investigate the properties and electronic structure of tetrahedrally coordinated zinc.

F. A. Cotton and J. S. Wood, *Inorganic Chemistry*, **3**, 245 (1963).

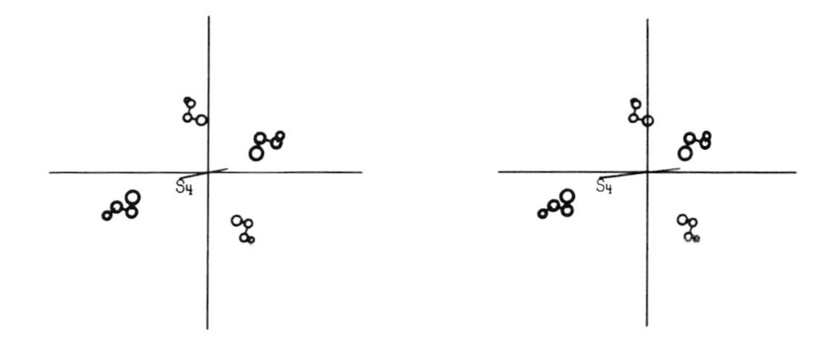
S4
S4

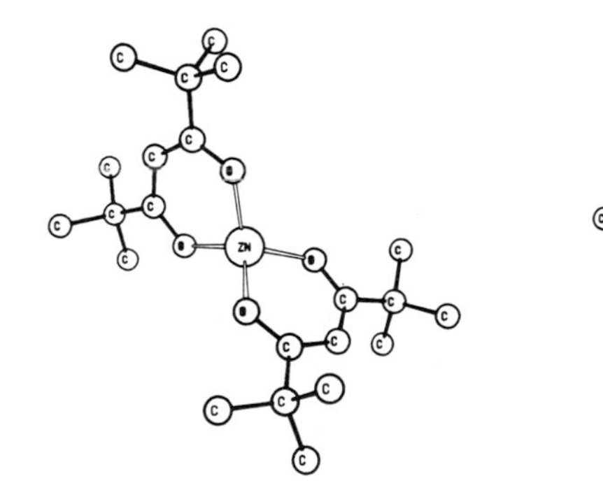
ZN

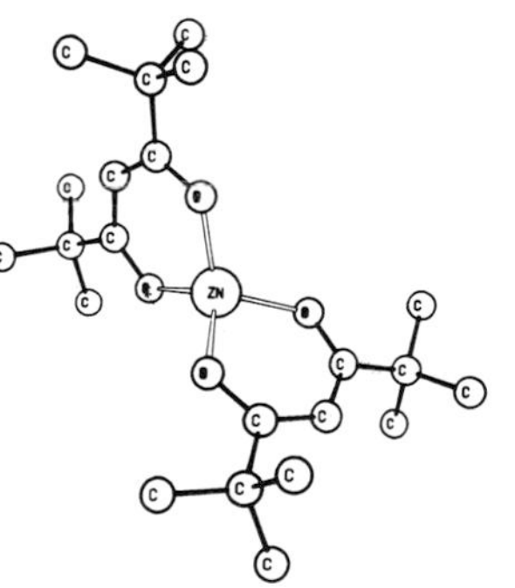
ZN

S_6 $\bar{3}$

A 6-fold rotation-reflection axis generates this point group. There are thus six symmetry operations in the point group. The application of the rotation–reflection operation either two or four times has the same effect as the application of the 3-fold axis operation. Thus the group may be thought of as being generated by adding an inversion center to point group C_3, hence the alternative notation C_{3i}. The three objects above the x, y plane are right-handed and the three objects below the plane are left-handed.

Order of group: 6

Crystallographic symmetry: Trigonal

Cerium magnesium nitrate hydrate $Ce_2Mg_3(NO_3)_{12} \cdot 24H_2O$

Only the $Ce(NO_3)_6$ part of this infinite **ionic crystal** is shown. In this ionic structure, the environment of the Ce ion has S_6 symmetry. Six nitrate groups are bonded to the Ce ion through two oxygen atoms. The salt has been extensively used in magnetic studies, in very low temperature cooling studies, and in spectroscopic studies. Proper interpretation of the results of these studies required precise structural information.

A. Zalkin, J. S. Forrester, and D. H. Templeton, *Journal of Chemical Physics*, **39**, 2881 (1963).

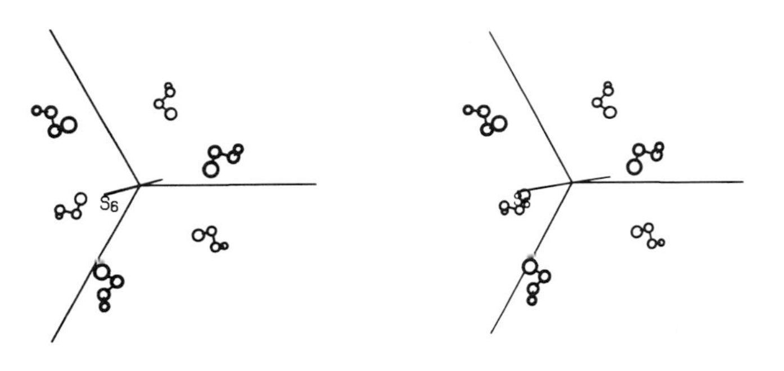
S6

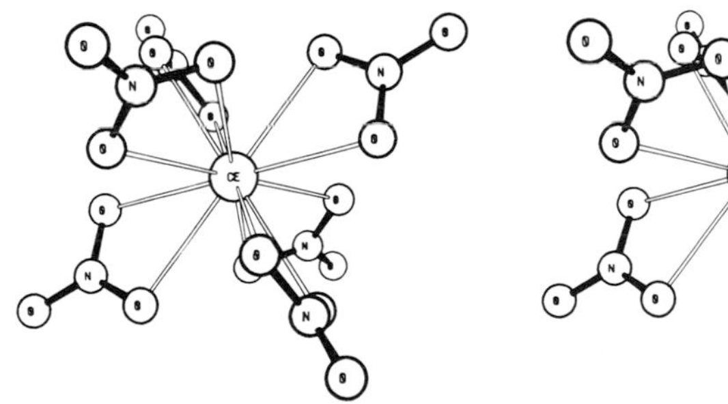
CE
N

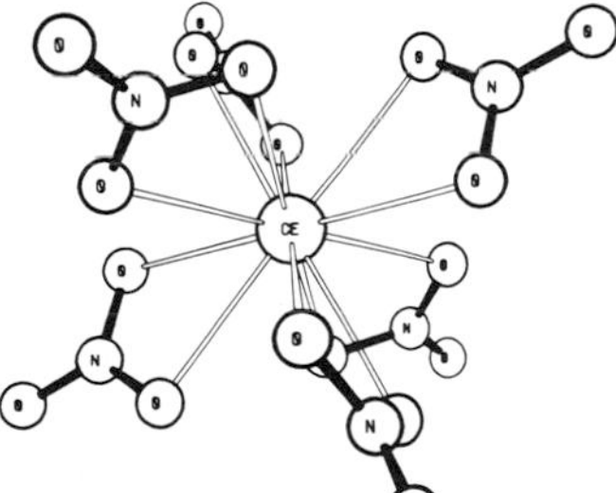
CE
N

Point Groups with a Single Rotation Axis and a Mirror Plane Perpendicular to the Axis

C_{nh} n/m

These point groups contain a rotation symmetry axis with a mirror plane perpendicular to the axis. Since the rotation axis is usually considered to be vertical, the mirror plane is horizontal and, in the Schoenflies notation, is denoted by h. Thus the symbols for these groups are

$$C_{nh}$$

The alternate Hermann-Mauguin notation is

$$n/m$$

which indicates a mirror plane perpendicular to the n-fold axis. The group C_{1h} or m is often denoted as C_s. It could also be called S_1. The group C_{3h} could be called S_3 or, in the Hermann-Mauguin notation, either $3/m$ or $\overline{6}$, although the former is preferred. Because a mirror plane is present, molecules with n/m symmetry are not optically active.

Schoenflies Symbol	Hermann-Mauguin Symbol
C_{1h}, C_s	m
C_{2h}	$2/m$
C_{3h}, S_3	$3/m$, $\overline{6}$
C_{4h}	$4/m$
C_{5h}	$5/m$
C_{6h}	$6/m$

C_{1h} *m*

The two symmetry elements are the identity element and a mirror plane, which in the figure is the horizontal plane. The upper object is obtained from the lower by a reflection in the mirror. One of the objects is right-handed and one is left-handed.

Order of group: 2

Crystallographic symmetry: Monoclinic

μ-Disulfidodiiron hexacarbonyl
$S_2Fe_2(CO)_6$

In a mixed crystal, which contains this species as well as $S_2Fe_3(CO)_9$, this molecule has symmetry *m* only. The approximate symmetry is higher; the reader should try to identify these approximate symmetry elements. Visualize the molecule as consisting of a disulfide (S_2) group bridging two identical iron tricarbonyl ($Fe(CO)_3$) fragments.

C. S. Wei and L. F. Dahl, *Inorganic Chemistry*, **4**, 493 (1965).

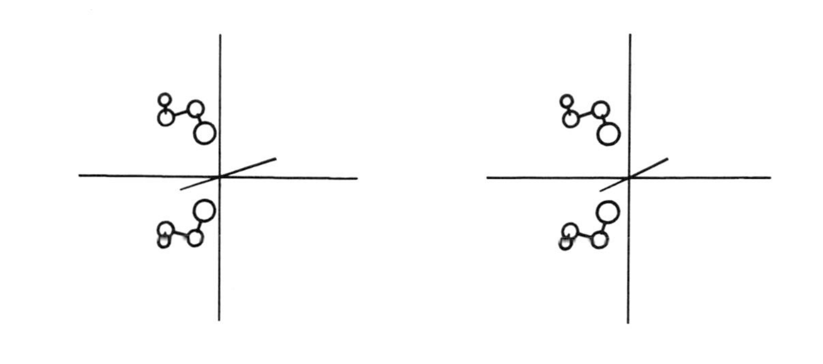

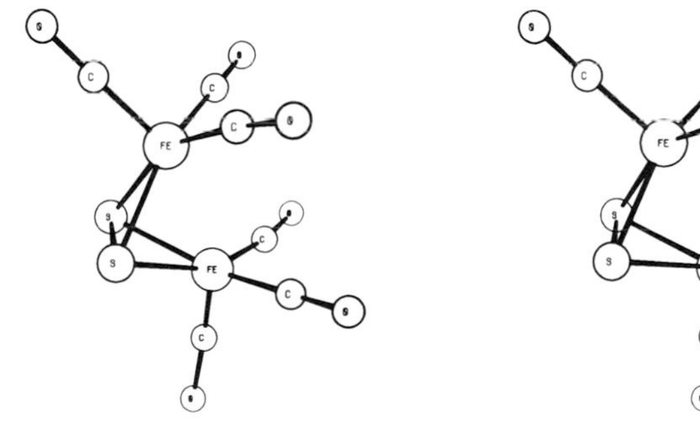

FE
S
C

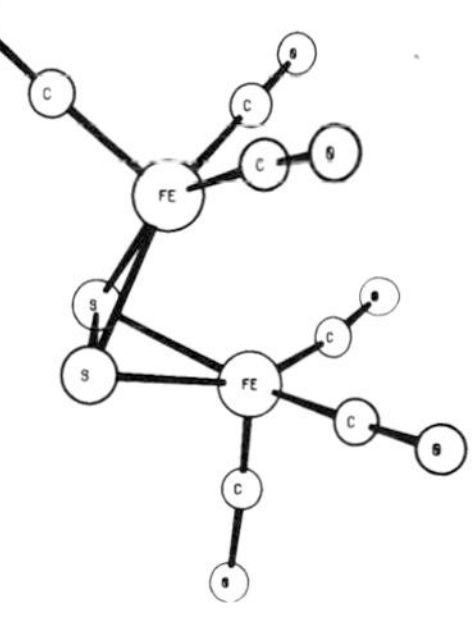

FE
S
C

C_{2h} 2/m

This point group is generated by two operations, the vertical 2-fold axis and the horizontal mirror plane. The combination of the two operations results in the third nonidentity symmetry operation of the point group: the inversion center. Both right-handed and left-handed objects appear.

Order of group: 4

Crystallographic symmetry: Monoclinic

Bis(tricarbonyl ruthenium(II) dibromide) $((CO)_3RuBr_2)_2$

The elemental formulation suggested that Ru was **pentacoordinated** by three carbonyls (CO) and two Br atoms. The x-ray analysis showed that the molecule is **dimeric** with **octahedral** coordination around each Ru.

S. Merlino and G. Montagnoli, *Acta Crystallographica*, **B24**, 424 (1968).

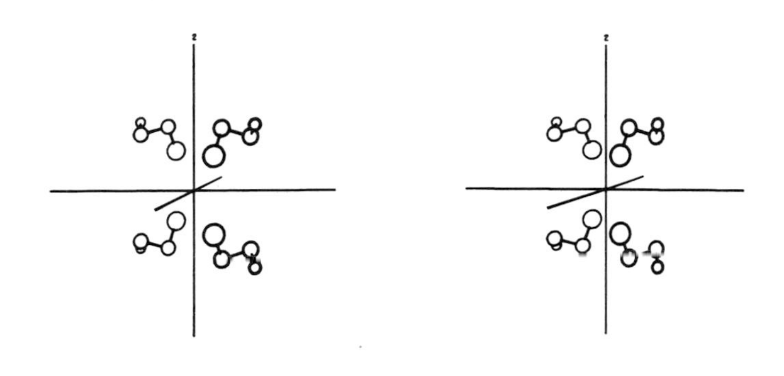

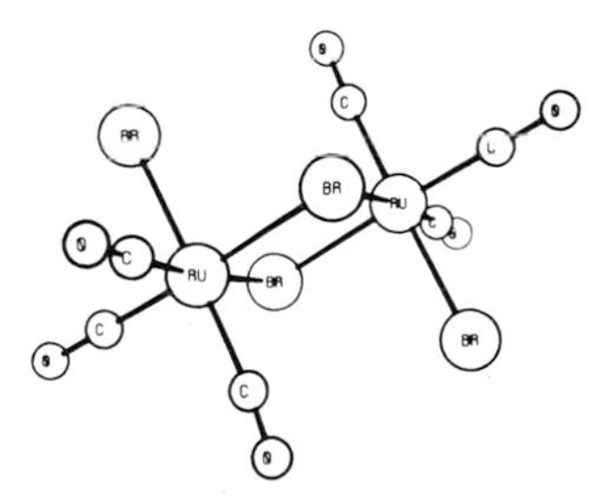

BR
RU
C
BR
RU
C
C
C
C
BR

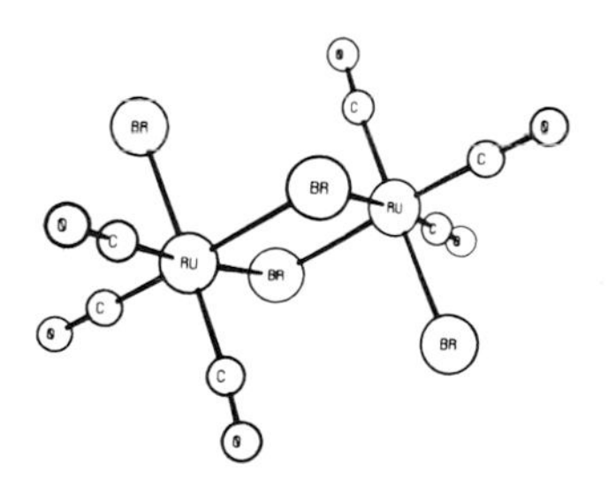

BR
BR
RU
C
C
RU
C
BR
C
C
BR

C_{3h} 3/m

The addition of a horizontal mirror plane to the point group C_3 generates this point group. It is identical to the group produced by the operations of a 3-fold rotation–reflection axis. There are both right- and left-handed objects, but there is no center of symmetry.

Order of group: 6

Crystallographic symmetry: Hexagonal

Tris(dithioglyoximato) molybdenum(0) $Mo(S_2C_2H_2)_3$

This is another illustration of a molecule that approximates a greater symmetry than is apparent in the crystal structure, which in this crystal is C_{3h}. The planes defined by the $H_2C_2S_2$ fragments do not include the molybdenum atom. What is interesting in this molecule is that the **coordination** around the metal atom is not octahedral. Instead, the six ligand sulfur atoms are arranged at the corners of a **trigonal prism**.

A. E. Smith, G. N. Schrauzer, V. P. Mayweg, and W. Heinreich, *Journal of the American Chemical Society*, **87**, 5798 (1965).

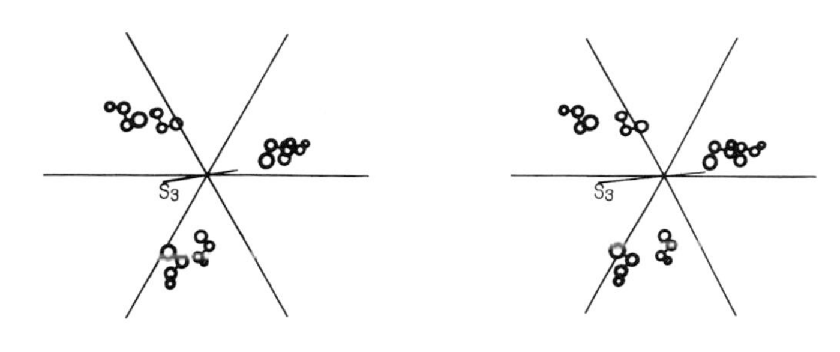
S3
S3

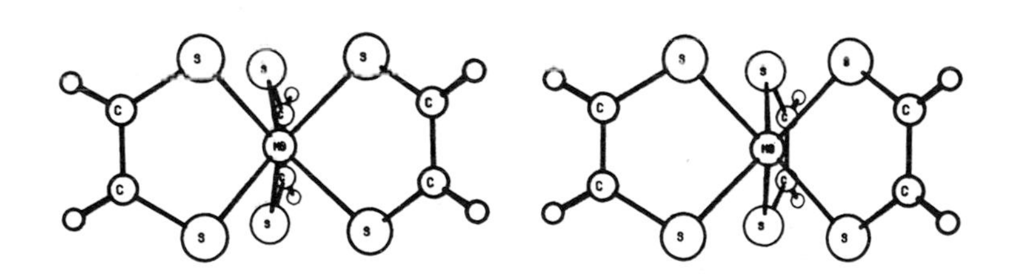

C_{4h} 4/m

This point group is obtained from C_4 by adding a mirror plane perpendicular to the 4-fold axis. The resulting point group has an inversion center. The four objects above the x, y plane are of one handedness, those below are of the other.

Order of group: 8

Crystallographic symmetry: Tetragonal

Tetramethylcyclobutadiene
$(CH_3)_4C_4$

Here is another possible configuration for the compound that is also pictured with point group C_4 (see page 74). In the point group illustrated here, one hydrogen atom from each methyl group lies in the plane of the ring.

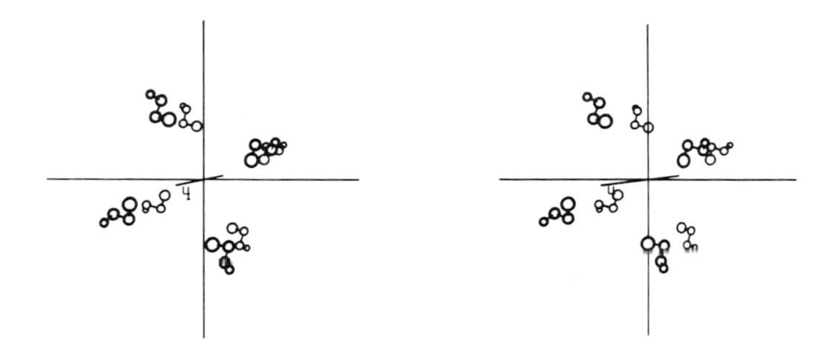

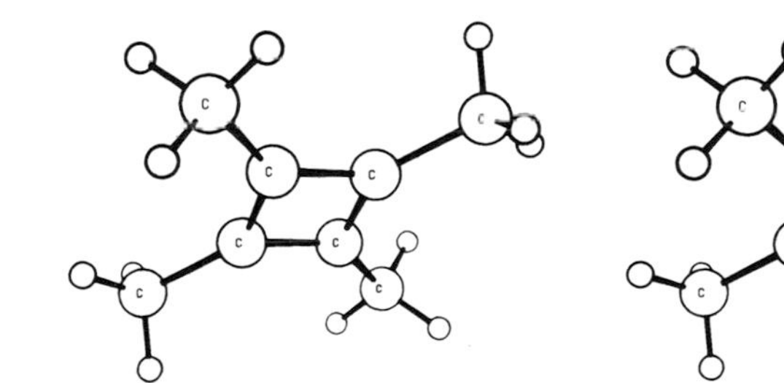

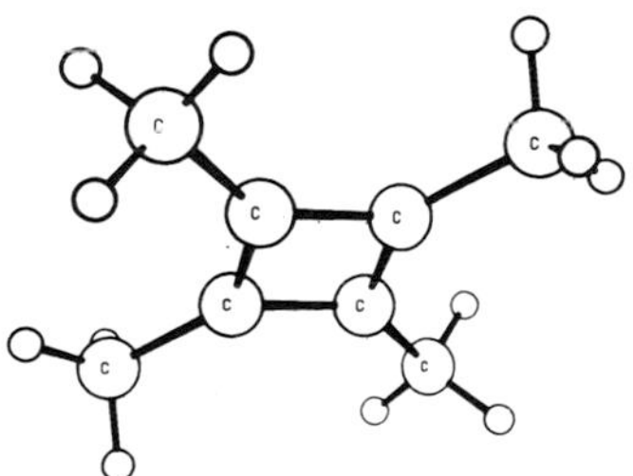

C_{5h} 5/m

This point group is obtained from point group C_5 by adding a mirror plane perpendicular to the 5-fold axis. The group elements are 5-fold rotations, 5-fold rotation-reflections, and the horizontal mirror plane. Again there is no center of symmetry. There are, however, molecules of both chiralities.

Order of group: 10

Crystallographic symmetry: none

Pentamethyldifluoroiodine
$(CH_3)_5F_2I$

The molecule has not been synthesized; it is analogous to the well characterized IF_7, the structure of which is still controversial although of approximately D_{5h} symmetry. Here, two fluorines and five methyl groups are bound to the central iodine atom. Again, the methyl groups are mirrored in the principal molecular plane.

R. E. La Villa and S. H. Bauer, *Journal of Chemical Physics*, **33**, 182 (1960).

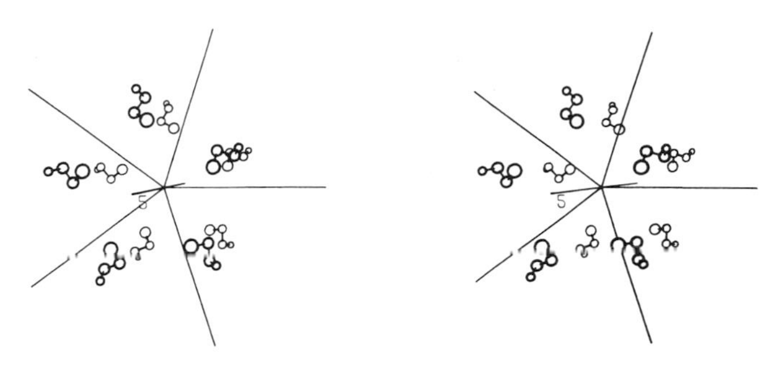
5

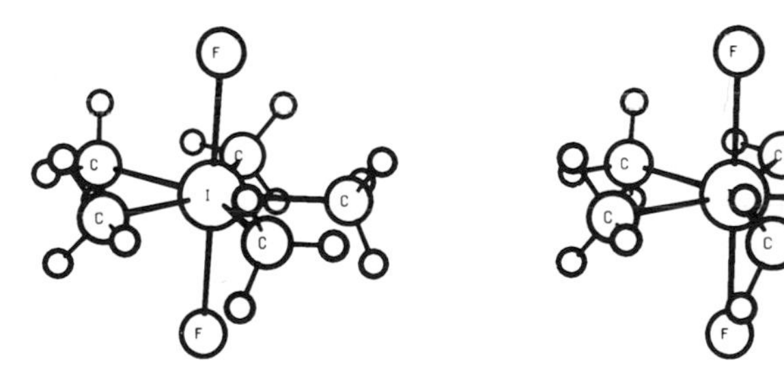
F
C
I

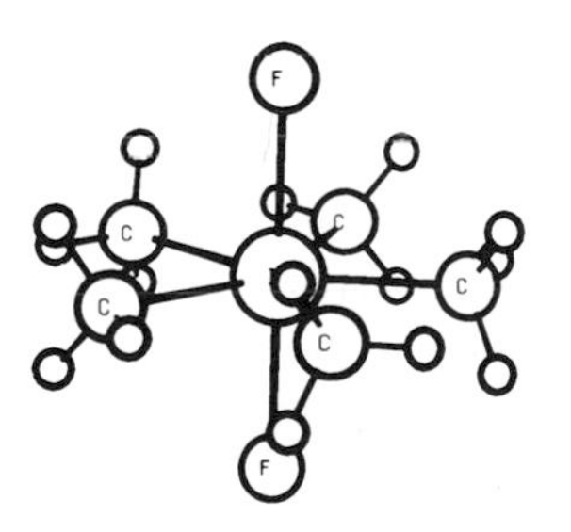

C_{6h} 6/m

Addition of a horizontal mirror plane to the point group C_6 generates this point group. The operations of the group include 6-fold rotations, the mirror plane, a 6-fold rotation–reflection axis, and finally, a center of symmetry. There are six left-handed and six right-handed objects.

Order of group: 12

Crystallographic symmetry: Hexagonal

Hexamethylbenzene
$(CH_3)_6C_6$

Another possible symmetry for the molecule that illustrated point group C_6 is C_{6h}. Again, just as C_6 was not the actual configuration attained in any one of the three solid phases, neither is C_{6h}. The actual symmetry in the solid state is approximately D_{3d} with distortions to $\overline{3}$. In C_{6h}, the methyl groups are all in the same orientation, and each has one hydrogen atom in the plane of the ring.

W. C. Hamilton, J. Edmonds, A. Tippe, and J. J. Rush, *Discussions of the Faraday Society*, **48**, 192 (1969).

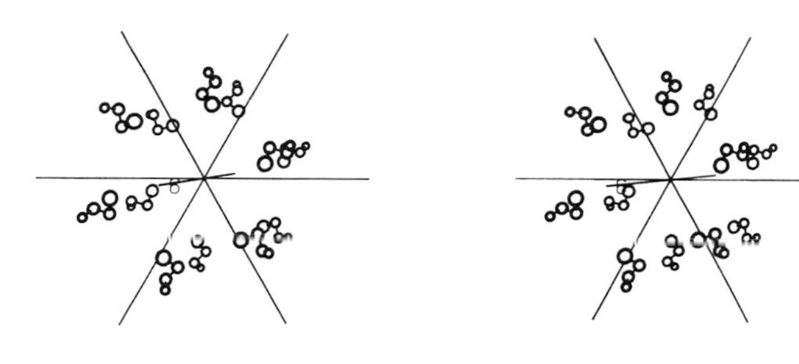
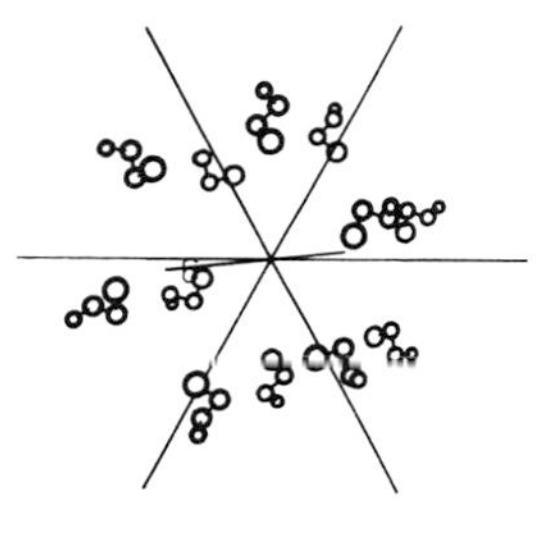

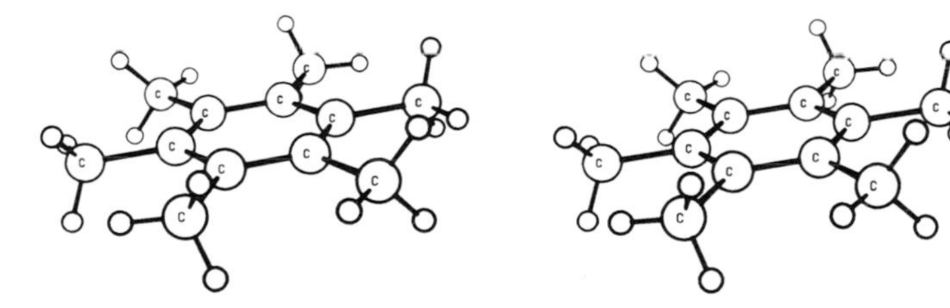
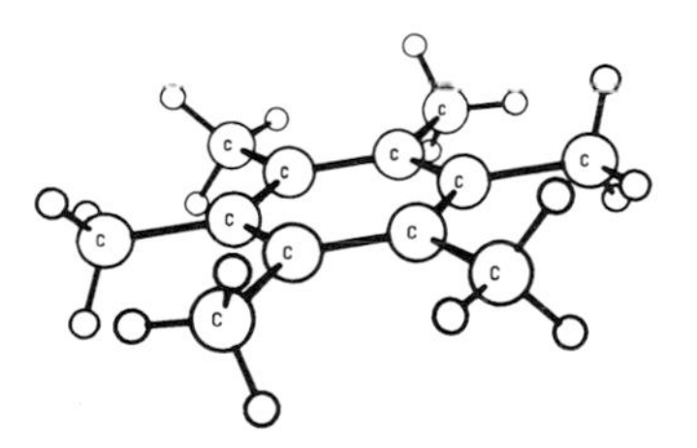

Simple Dihedral Point Groups

D_n $n2-$

In these groups, the principal rotation axis of order n has perpendicular to it n 2-fold axes. These groups are called *dihedral* and have the Schoenflies symbol

$$D_n$$

The alternative Hermann-Mauguin notation has as its principal symbol the integer representing the order of the axis. This is followed by 2 to denote the 2-fold axes. If n is even, a third digit 2 is added to indicate that a second set of 2-fold axes midway between the first is generated by the basic symmetry operations. Thus, the group 422 has 2-fold axes separated by 45°.

Molecules possessing D_n symmetry are optically active.

Schoenflies Symbol	Hermann-Mauguin Symbol
D_2	222
D_3	32
D_4	422
D_5	52
D_6	622

D_2 222

This point group contains four operations, the identity operation and three perpendicular 2-fold axes. These axes are indicated in the figure as coincident with the x, y, and z coordinate axes. A rotation around the vertical z axis followed by a rotation around the horizontal x axis is exactly equivalent to a single rotation around the horizontal y axis. The objects in this point group are of the same chirality.

Order of group: 4

Crystallographic symmetry: Orthorhombic

Zirconium(IV) acetylacetonate
$Zr(CH_3COCHCOCH_3)_4$

The acetylacetonate $(CH_3—CO—CH—CO—CH_3)^-$ ion, a beta-diketone, readily binds many metals through its two oxygens, thus acting as a **bidentate** ligand. A number of these complexes involve four ligands around the metal and a consequent eight-coordination of the metal atom. There are two possible idealized structures: the **dodecahedron**, and the **tetragonal antiprism** shown here. The hydrogen atoms are not illustrated.

J. V. Silverton and J. L. Hoard, *Inorganic Chemistry*, **2**, 243 (1963).

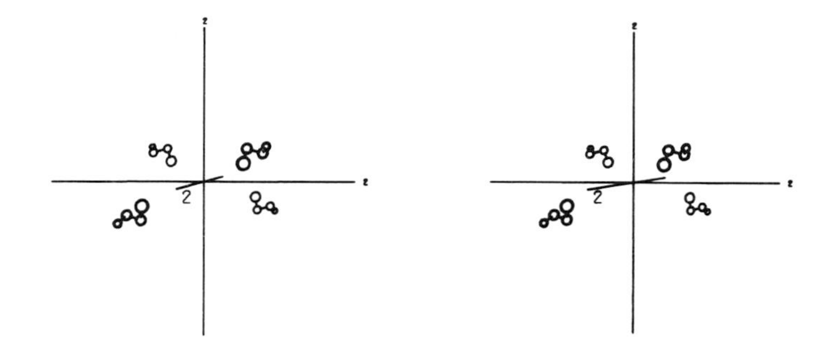
2
2

ZR

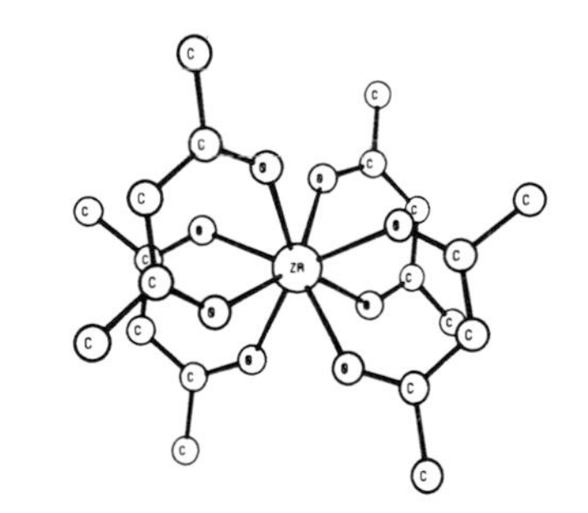
ZR

D_3 32

Perpendicular to the 3-fold rotation axis are three 2-fold axes. A rotation by $2\pi/3$ followed by a rotation around any one of the 2-fold axes is equivalent to a rotation about another of the 2-fold axes. All of the objects are of the same handedness.

Order of group: 6

Crystallographic symmetry: Trigonal

Tris(hexamethyldisilylamine) iron(III)
$\{[(CH_3)_3Si]_2N\}_3Fe$

The composition of this compound suggested that it contained an iron atom bonded to three nitrogen ligands; however, previous examples of potential three-bonded systems had proved to be **polymeric** species. An x-ray diffraction investigation confirmed that this was the first known example of a three-bonded iron compound. The hydrogen atoms on the methyl groups are not shown.

D. D. Bradley, M. B. Hursthouse, and P. F. Rodesiler, *Chemical Communications*, 14 (1969).

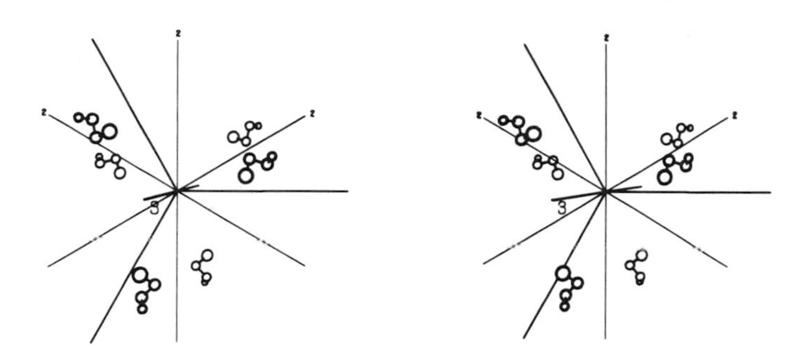

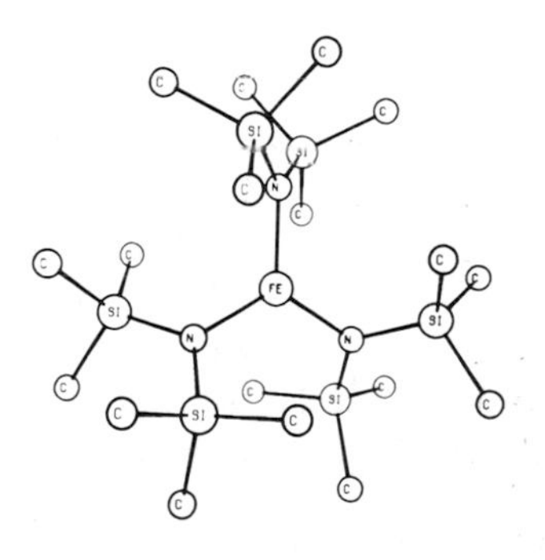
FE
N
SI
C

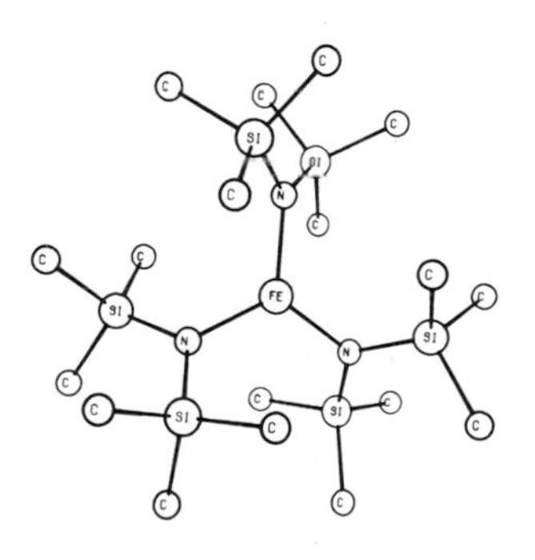
FE
N
SI
C

D_4 422

This group is generated by a 4-fold axis with four perpendicular 2-fold axes. All of the objects are of the same handedness.

Order of group: 8

Crystallographic symmetry: Tetragonal

Tetraaquodichlorocobalt(II)
$(H_2O)_4Cl_2Co(II)$

A typical octahedral complex of cobalt. Four water molecules are bonded to the cobalt along their 2-fold axes. The apical positions are occupied by chloride ions. The species has been observed in crystals of $CoCl_2 \cdot 6H_2O$, which is better formulated as $[Co(H_2O)_4Cl_2] \cdot 2H_2O$. Optical activity would seem to be indicated by the 422 symmetry of this species. But rotation of the water molecules around the 2-fold axes would easily convert left-handed ions into right-handed ions, and a solution containing ions of both orientations would not be optically active.

J. Mizuno, *Journal of the Physical Society of Japan*, **15**, 1412 (1960).

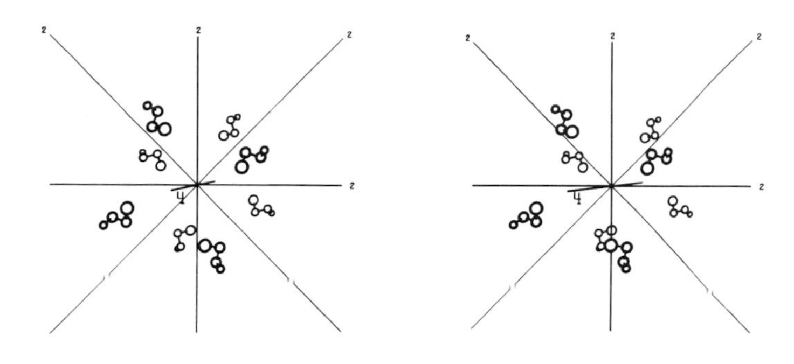

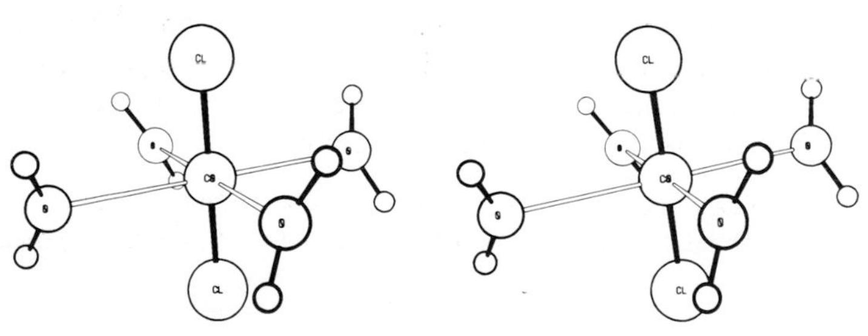

D_5 52

Point group D_5 is generated by a 5-fold axis in combination with five 2-fold axes perpendicular to it. All the objects are of the same handedness.

Order of group: 10

Crystallographic symmetry: none

Bis(cyclopentadienyl) iron(II)
$(C_5H_5)_2Fe$

Ferrocene was the first **pi-bonded sandwich** compound to be made. It is shown here in a configuration in which the cyclopentadienyl $(C_5H_5)^-$ rings are neither fully **eclipsed** nor fully staggered. In the crystal the symmetry is probably higher, namely D_{5d}.

T. J. Kealy and P. L. Pauson, *Nature*, **168**, 1039 (1951).
J. D. Dunitz, L. E. Orgel, and A. Rich, *Acta Crystallographica*, **9**, 373 (1956).

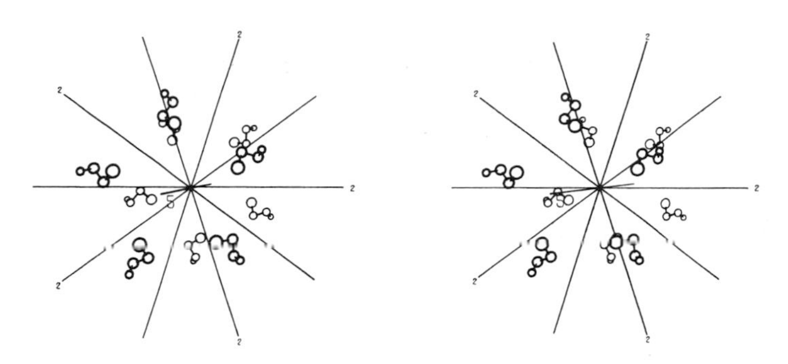
2
2
2
2
2
2
5

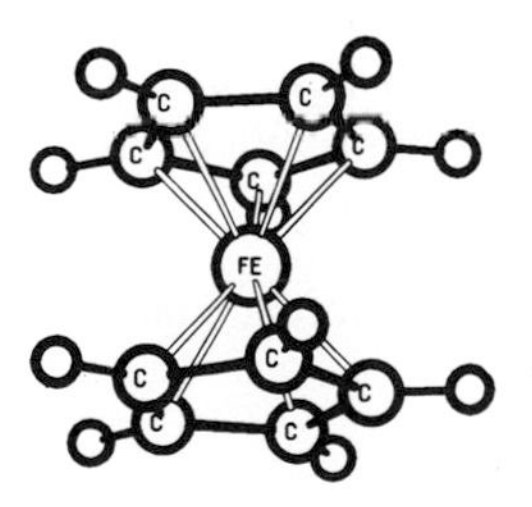
C
C
C
C
C
FE
C
C
C
C
C

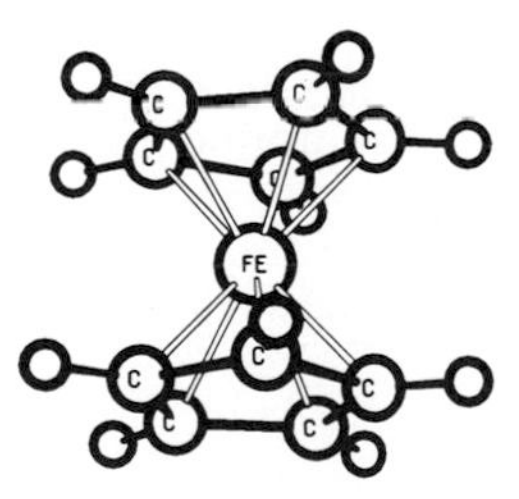
C
C
C
C
C
FE
C
C
C
C
C

D_6 622

This is a simple dihedral point group generated by a 6-fold axis with six perpendicular 2-fold axes 30° away from one another in the plane. All objects are of the same handedness.

Order of group: 12

Crystallographic symmetry: Hexagonal

Hexaminobenzene
$C_6(NH_2)_6$

The complete structure was not determined in an early x-ray diffraction study; the hydrogen atoms are shown here in a probable configuration with the **amino** groups rotated from the plane of the ring to avoid crowding.

J. E. Knaggs, *Proceedings of the Royal Society*, (London), **A131**, 612 (1931).

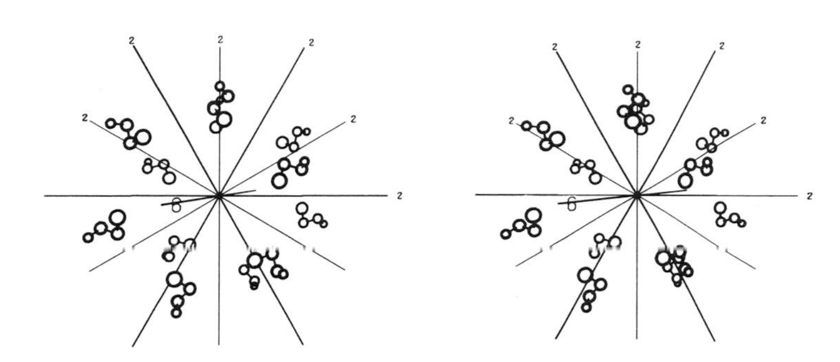

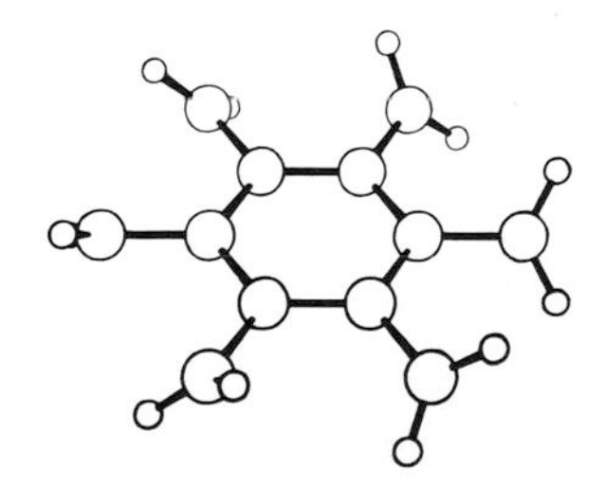

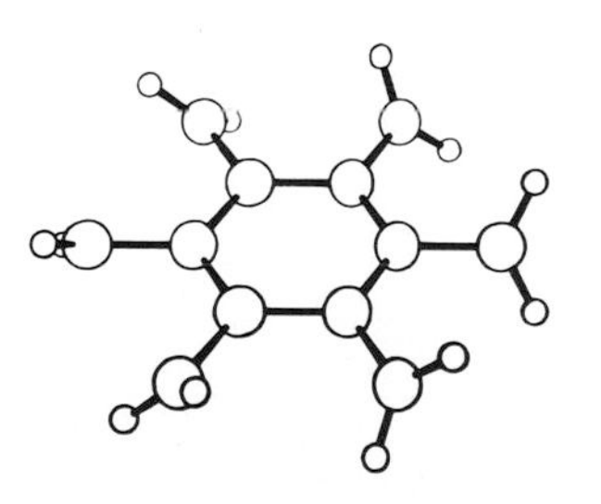

Dihedral Groups with Vertical Diagonal Mirror Planes

D_{nd} $\bar{n}2m$

These groups are derived from the groups D_n by adding vertical mirror planes that bisect the angles between the 2-fold axes. The subscript d is derived from the fact that such mirror planes are called *diagonal* mirror planes. These groups also contain rotation–inversion axes with mirror planes which include the rotation-inversion axis. The usual Hermann-Mauguin notation is based on the latter symmetry elements.

Schoenflies Symbol	Hermann-Mauguin Symbol
D_{2d}	$\bar{4}2m$
D_{3d}	$\bar{3}2m$*

*The 2 is often omitted.

D_{2d} $\bar{4}2m$

Adding 2-fold axes perpendicular to the 4-fold axis of the point group S_4 generates this group. Alternatively, the group may be derived from D_2 by adding vertical mirror planes that include one 2-fold axis but bisect the angles formed by the other two. The operations of the group may be described in terms of two mirror operations, three 2-fold rotations, two rotation-reflections, and the identity. There is no inversion center. Both right- and left-handed objects exist, and optical activity is not a characteristic of molecules possessing the symmetry of this group or any other D_{nd} group.

Order of group: 8

Crystallographic symmetry: Tetragonal

[Tetramethyl-tetra(tetracarbonyl iron)] tritin $(CH_3)_4Sn_3Fe_4(CO)_{16}$

This molecule provides a fine example of a metallic **cluster compound**, in which four iron atoms are bonded to a central tin atom while, in turn, the irons are surrounded by other ligands. The structure shown, obtained by x-ray diffraction, is not that expected to result from the chemical preparation, which was carried out to obtain a more compact metallic cluster. The methyl (CH_3) and carbonyl (CO) groups are shown as single spheres.

R. M. Sweet, C. J. Fritchie, Jr., and R. A. Schunn, *Inorganic Chemistry*, **6**, 749 (1967).

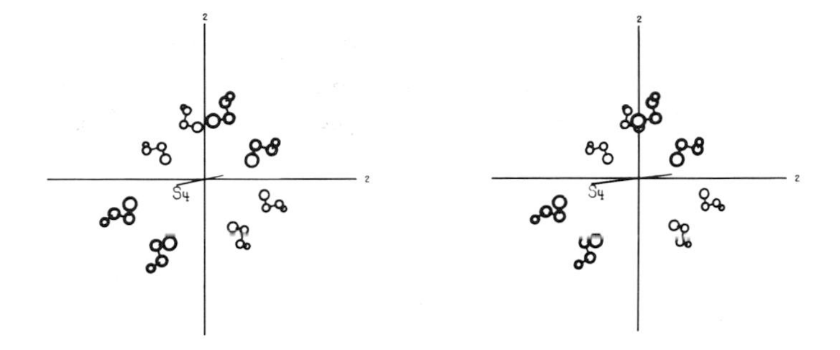

2
2
S_4
2
2
S_4

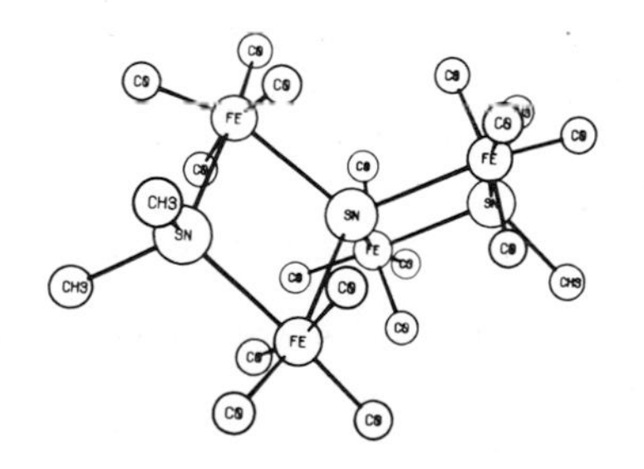

FE
SN
CH3
CO

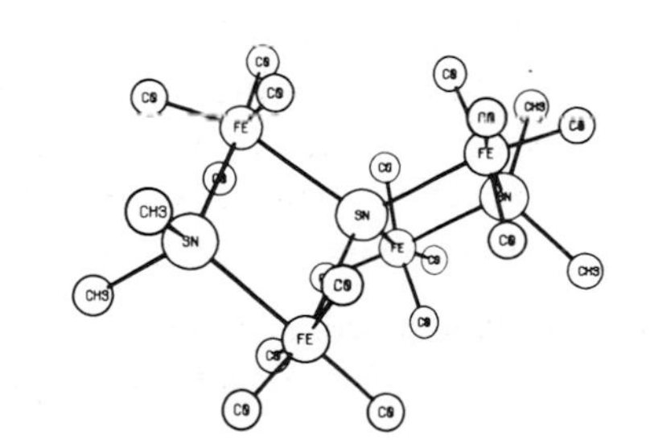

FE
SN
CH3
CO

D_{3d} $\bar{3}m$

This point group may be obtained from point group D_3 by adding an inversion center. Diagonal mirror planes that bisect the angles between the 2-fold axes are generated. Objects of both hands exist.

Order of group: 12

Crystallographic symmetry: Trigonal

Bis(ethylthio)nickel(II) hexamer
$(C_2H_5)_{12}S_{12}Ni_6$

This **hexamer** is one of several compounds that may be obtained from reaction of diethyl-disulfide (C_2H_5—S—S—C_2H_5) with nickel carbonyl, $Ni(CO)_4$. The crown-shaped molecule is one of extraordinary beauty. The ethyl groups (C_2H_5) are denoted by Et and the **beta-carbon** atoms do not share the full symmetry of the rest of the molecule.

P. Woodward, L. F. Dahl, E. W. Abel, and B. C. Crosse, *Journal of the American Chemical Society*, **87**, 5251 (1965).

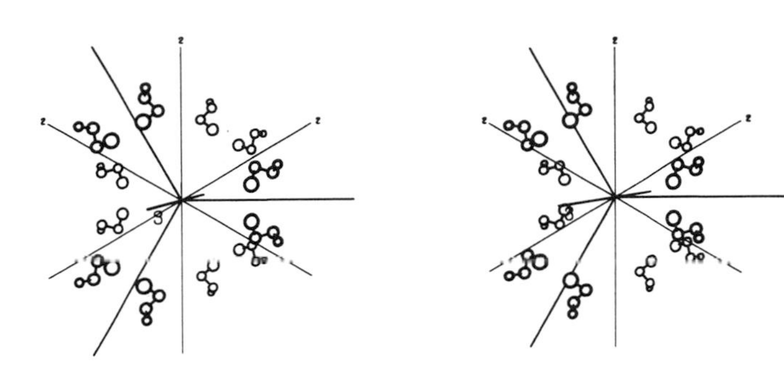

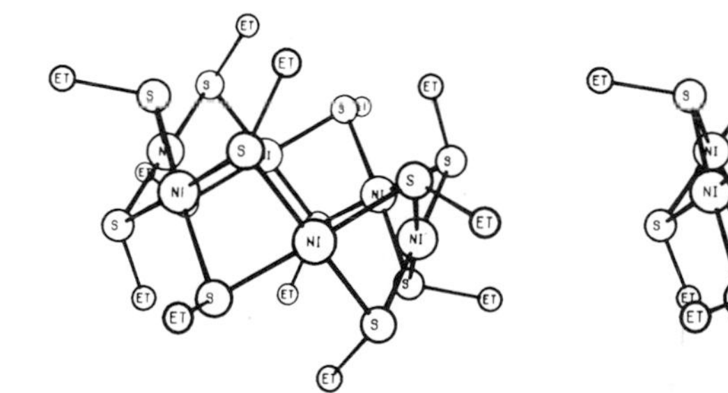
ET
S
NI

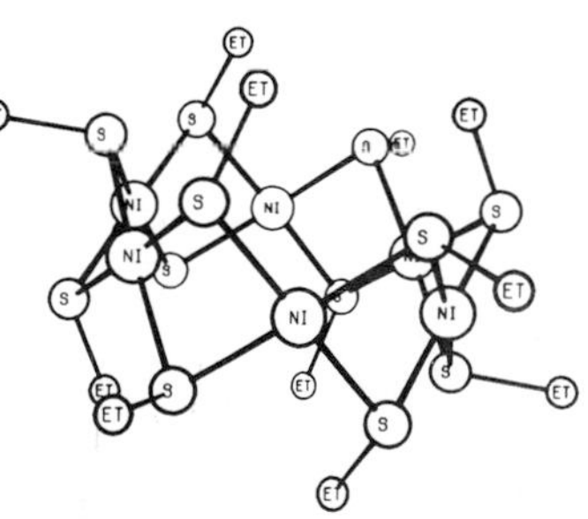
ET
S
NI

Dihedral Groups with Horizontal Mirror Planes

D_{nh} n/mm–

These groups are obtained from the simple dihedral groups D_n by adding mirror planes perpendicular to the principal axis. Thus the Schoenflies notation is

$$D_{nh}$$

The alternate notation cites an axis with a perpendicular mirror plane, that is,

$$n/m$$

followed by the notation that there are n mirror planes that also include the principal axis and, for the even rotation axes, a second set of mirror planes which bisect the angles formed by the first set. For the odd rotation axes, there is only one set of equivalent mirror planes. Thus we have n/mmm for n even and n/mm for n odd. The equally valid notation $\bar{6}m2$ for $3/mm$ is usually used by crystallographers.

Schoenflies Symbol	Hermann-Mauguin Symbol
D_{2h}	mmm, $2/mmm$
D_{3h}	$3/mm$, $\bar{6}m2$
D_{4h}	$4/mmm$
D_{5h}	$5/mm$
D_{6h}	$6/mmm$

D_{2h} mmm

This point group has eight symmetry elements. It may be obtained from the point group D_2 by adding a horizontal mirror plane, which in turn generates the perpendicular mirror planes. There are three mirror planes, each defined by two of the three coordinate axes. Each of the three coordinate axes is also a 2-fold axis, since it is the intersection of two mirror planes. The eight symmetry operations are thus the identity element, the center of symmetry, the three 2-fold rotations, and the three mirror planes. The point group contains both right- and left-handed objects. No optical activity is possible with molecules of D_{nh} symmetry.

Order of group: 8

Crystallographic symmetry: Orthorhombic

Bis(O,O′-diethyldithiophosphato) nickel(II) adduct with pyridine $[(C_2H_5O)_2PS_2]_2(C_5H_5N)_2Ni$

The **adducts** formed by metal compounds in solution are sometimes useful in analytical chemistry; the nature of these adducts is often a subject of controversy. The crystal and molecular structures were determined by x-ray diffraction in order to establish unequivocally the geometry of the product in this adduct. It was of particular importance to determine if the two pyridines (C_5H_5N) were bound to the nickel. Such proved to be the case. (The hydrogen atoms are not shown in the figure).

S. Ooi and Q. Fernando, *Inorganic Chemistry*, **6**, 1558 (1967).

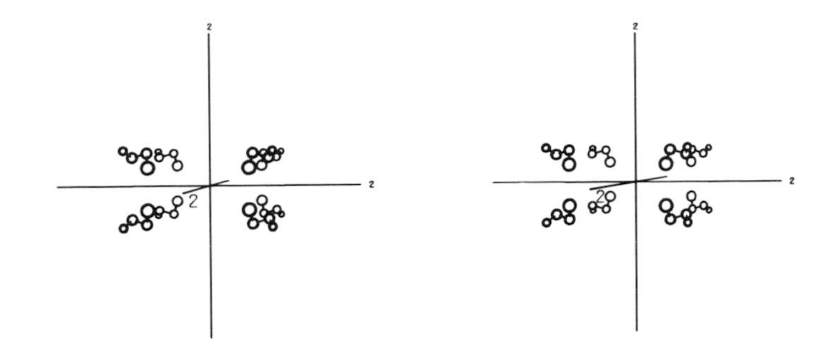

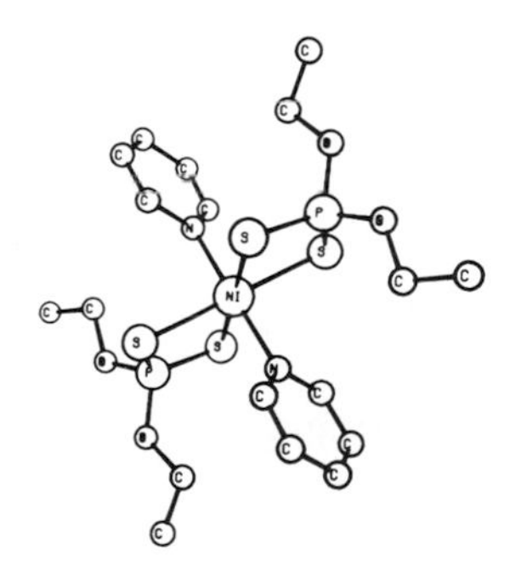

NI
S
P
N
C

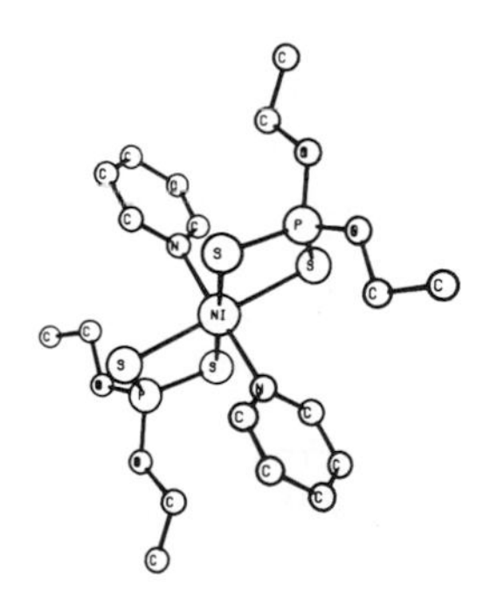

NI
S
P
N
C

D_{3h} 3/mm

Adding a mirror plane perpendicular to the 3-fold axis of the point group D_3 generates this group. There are three 2-fold axes in the plane, and there are vertical mirror planes as well as a single horizontal mirror plane. Both right- and left-handed molecules exist.

Order of group: 12

Crystallographic symmetry: Hexagonal

Cesium dodecachlorotrirhenate(III)
$Cs_3(ReCl_4)_3$

This compound was traditionally formulated as containing Cs^+ and $ReCl_4^-$. Certain properties, however, indicated that the structure was not that simple. An x-ray diffraction study demonstrated that it was a **trimer** containing the $Re_3Cl_{12}^{-3}$ ion shown here, with D_{3h} symmetry. Like the molecule that illustrated the group D_{2d}, this is an example of a metallic cluster compound.

J. A. Bertrand, F. A. Cotton, and W. A. Dollase, *Inorganic Chemistry*, **2**, 1166 (1963).

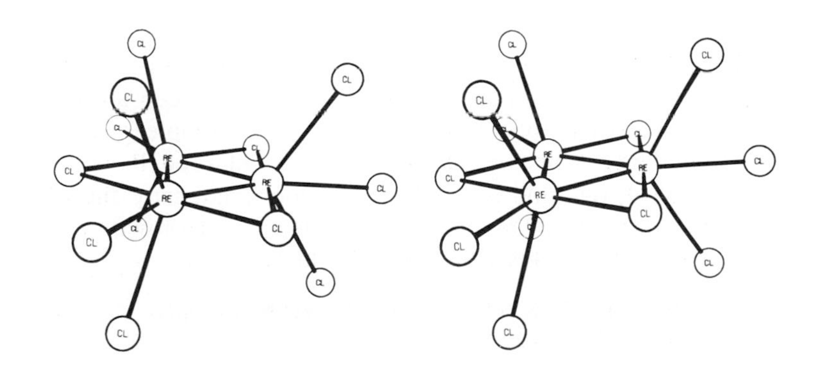
CL
RE

D_{4h} 4/mmm

The point group may be obtained from D_4 by adding a horizontal mirror plane that is perpendicular to the 4-fold axis. Symmetry operations include 2-fold and 4-fold axes, horizontal and vertical mirror planes, as well as 4-fold rotation–reflection operations and an inversion center. Mirror planes coincide with every pair of coordinate axes, also lie halfway between the x and y axis, and include the z axis. Both right- and left-handed objects exist.

Order of group: 16

Crystallographic symmetry: Tetragonal

Zinc tetraphenylporphine dihydrate $Zn[(C_6H_5)_4C_{20}H_7N_4]\cdot 2H_2O$

Porphine complexes of metals are typical of a large class of compounds in which a ring system with four interior nitrogen atoms binds a metal atom in a square planar arrangement. Such complexes exist as fragments of the large molecules **hemoglobin** and **chlorophyll**. In the structure of this **dihydrate**, zinc is octahedrally coordinated by the addition of two water molecules. The hydrogen atoms are not shown; actually, those on the water molecules violate the 4-fold symmetry.

E. B. Fleischer, C. K. Miller, and L. E. Webb, *Journal of the American Chemical Society*, **86**, 2342 (1964).

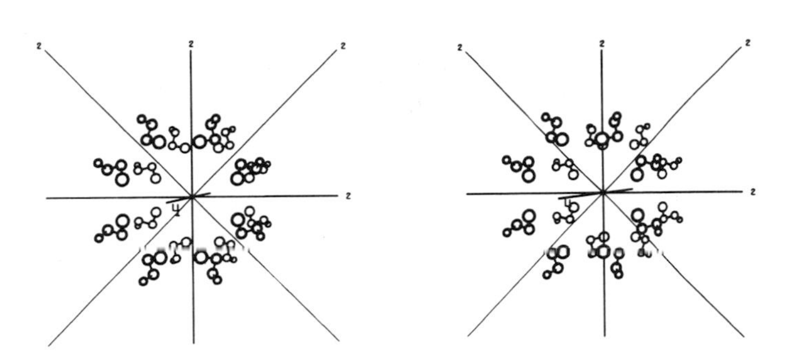

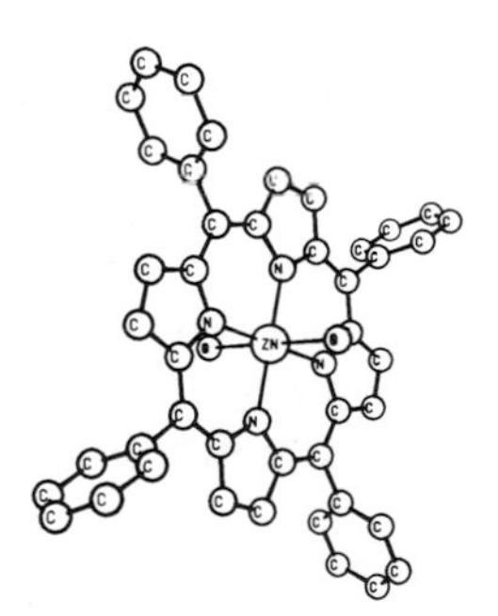

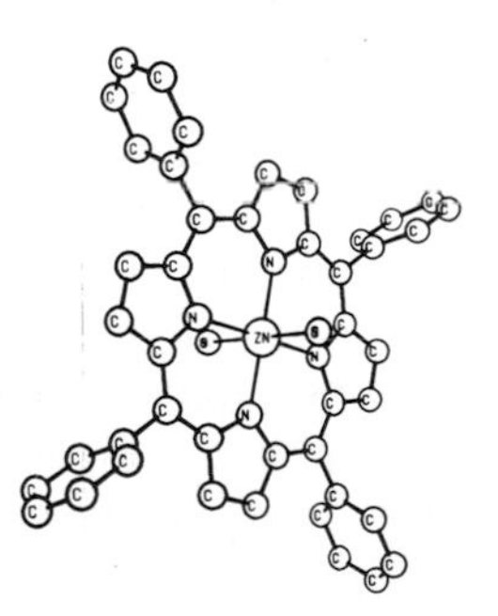

D_{5h} 5/mm

This group is obtained from the point group D_5 by adding a mirror plane perpendicular to the 5-fold axis. There are five 2-fold axes, five vertical mirror planes, and 5-fold rotation-reflection axes.

Order of group: 20

Crystallographic symmetry: none

Protoactinium pentachloride
$PaCl_5$

Although the compound of this composition was first synthesized in 1934, sufficient quantities for a structural investigation were not available until three decades later. The structure was found to be one in which the protoactinium atoms have 7-fold coordination; the **stoichiometry** is achieved by sharing chlorine atoms between two protoactinium atoms. The **pentagonal bipyramidal** environment of the protoactinium atom is shown in the drawing. The actual geometry in the crystal deviates slightly from the ideal symmetry.

R. P. Dodge, G. S. Smith, Q. Johnson, and R. E. Elson, *Acta Crystallographica*, **22**, 85 (1967).

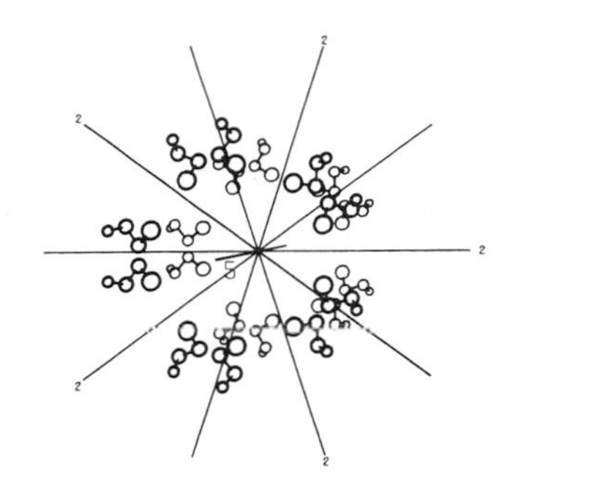
2
2
2
2
2

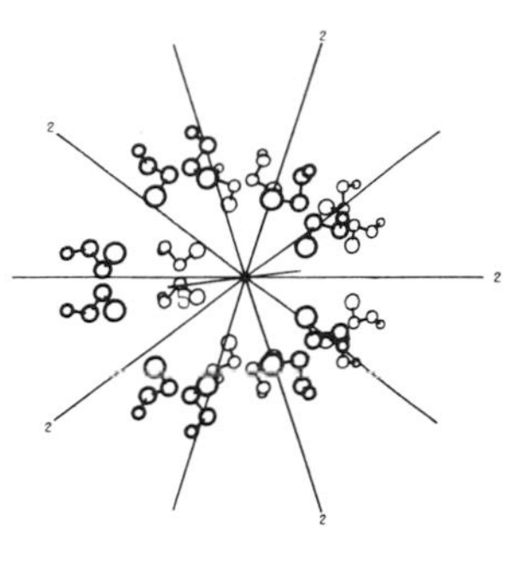
2
2
2
2
2

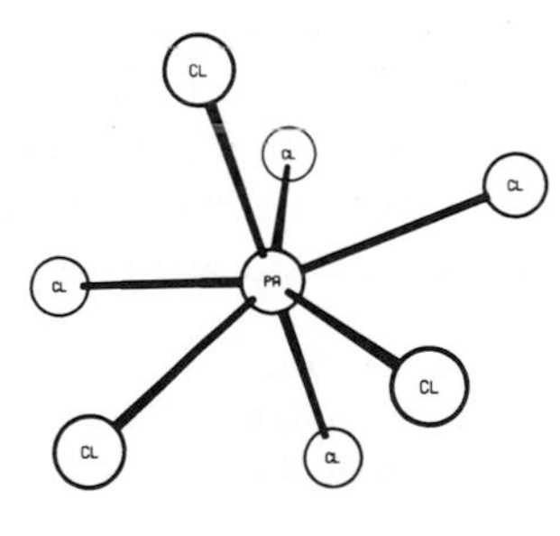
CL
CL
CL
CL
PR
CL
CL
CL

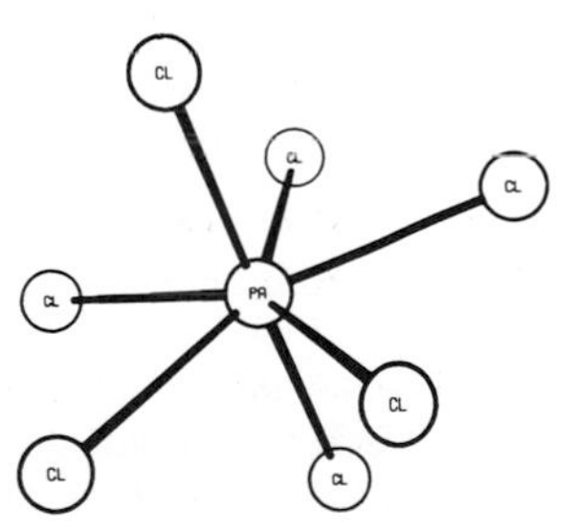
CL
CL
CL
CL
PR
CL
CL
CL

D_{6h} 6/mmm

The addition of a mirror plane perpendicular to the 6-fold axis of D_6 generates this group. There are six mirror planes that include the 6-fold axis. Three of these include the 2-fold axes; three bisect the angles between the 2-fold axes. There is a center of symmetry, and objects of both hands exist.

Order of group: 24

Crystallographic symmetry: Hexagonal

Dibenzene chromium
$(C_6H_6)_2Cr$

The symmetry of this molecule, which consists of two benzene molecules bound to a chromium atom equidistant from both, has been disputed for some years. At one time, it seemed likely that it did not have a 6-fold axis and that the bond lengths in the benzene rings alternated in length. But the most recent evidence confirms the D_{6h} structure shown here.

F. Jellinek, *Journal of Organometallic Chemistry*, **1**, 43 (1963).
E. O. Fischer, *Journal of Organometallic Chemistry*, **7**, 121 (1967).
E. Keulen, *Dissertation*, Groningen University, The Netherlands (1967).

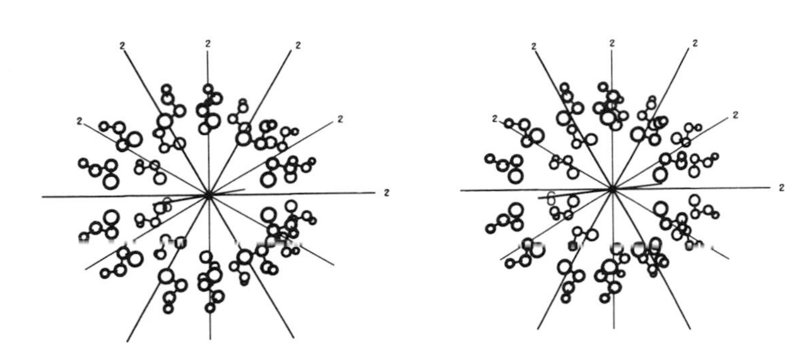
2
2
2
2
2
2
2
2
2
2
2
2

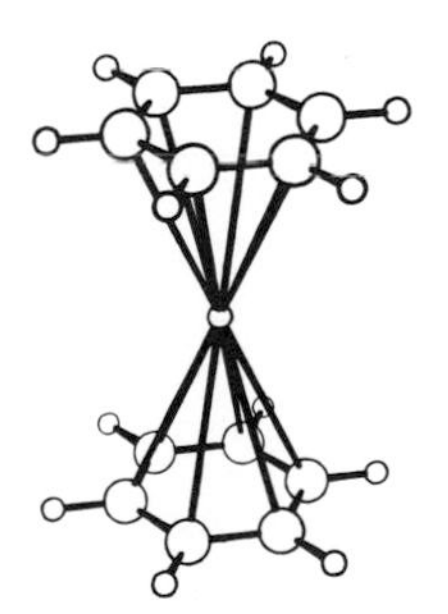

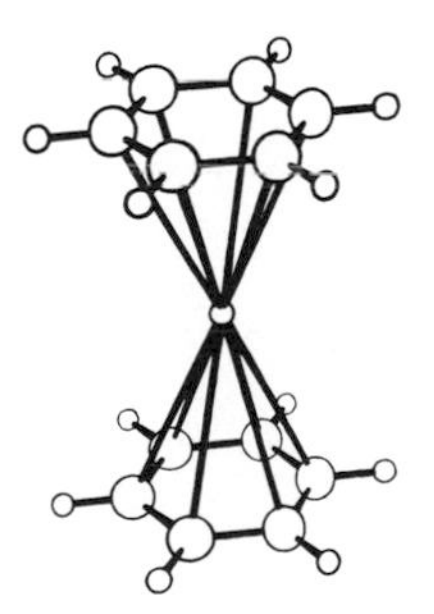

The Cubic Point Groups

T 23–O

The cubic point groups have the characteristic feature of having three perpendicular 2-fold axes (as in group 222) with a 3-fold axis equidistant from the three 2-fold axes, that is, along the diagonal of a cube formed by the three 2-fold axes. One of the cubic groups has the symmetry of the regular tetrahedron, one that of the regular octahedron. The Schoenflies symbols for the cubic groups thus are based on the symbols T (for tetrahedral) and O (for octahedral). The subscripts h and d have the same meanings as in simpler point groups. In the Hermann-Mauguin notation, the 2-fold or 4-fold axis is listed first, then the 3-fold axis, and finally, any symmetry axis directed along a cube face diagonal or a mirror plane perpendicular to such a diagonal.

Schoenflies Symbol	Hermann-Mauguin Symbol
T	23
T_d	$\bar{4}3m$
O	432
O_h	$4/m\bar{3}2/m$*
T_h	$2/m\bar{3}$*

*The Hermann-Mauguin symbols for O_h and T_h are often abbreviated to $m3m$ and $m3$.

T 23

Of all cubic point groups, this is the simplest. It may be obtained from the point group 222 by adding 3-fold axes that are equidistant from the three 2-fold axes. The twelve symmetry elements include the rotations about the 3-fold axes, the rotations about the 2-fold axes, and the identity element. All objects are of the same handedness. A molecule with this symmetry would be optically active.

Order of group: 12

Crystallographic symmetry: Cubic

Neopentane (2,2-dimethylpropane)
$C(CH_3)_4$

If the spheres representing atoms in this molecule were drawn full-size, the molecule would be seen to be nearly spherical. It provides an example of a so-called plastic crystal because of the ability of these spheres to tumble rapidly in the crystal lattice.

A. H. Mones and B. Post, *Journal of Chemical Physics*, **20**, 755 (1952).

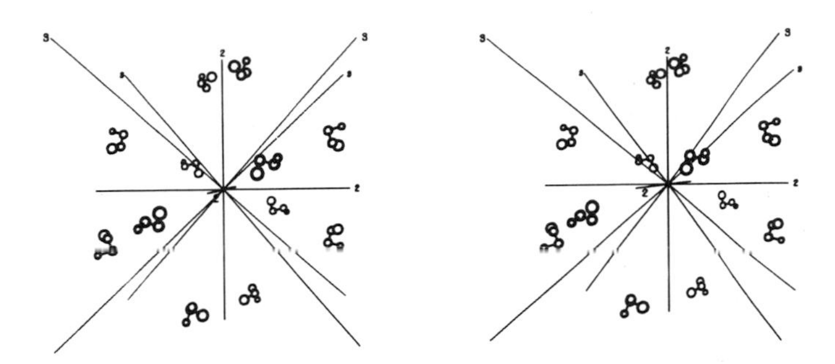

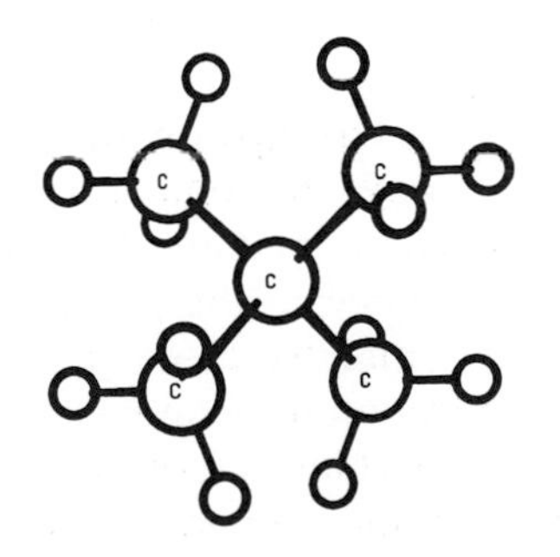
C
C
C
C
C

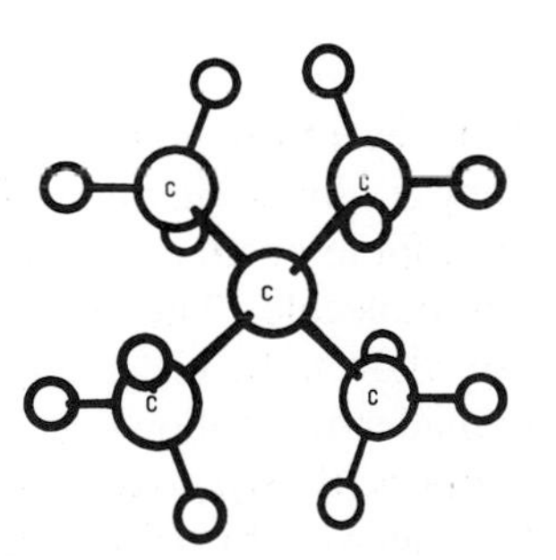
C
C
C
C
C

T_d $\bar{4}3m$

This point group is obtained from the point group T by adding mirror planes that contain the 3-fold and 2-fold axes. The symmetry elements thus include the identity, the operations of 3-fold and 2-fold rotation axes, mirror planes, and 4-fold rotation-reflection axes that are generated by a combination of other symmetry elements. The results of the S_4 axis are especially prominent in the figure. Both left- and right-handed objects exist, but there is no inversion center. A molecule with this symmetry would not be optically active.

Order of group: 24

Crystallographic symmetry: Cubic

μ_4-Oxo-hexa-μ-chloro-tetrakis [(triphenylphosphine oxide) copper(II)] $Cu_4OCl_6[(C_6H_5)_3PO]_4$

This beautiful compound contains a tetrahedron of pentacoordinated copper atoms surrounding a central oxygen atom. It was obtained in an attempt to produce a monomeric species containing only one copper atom per molecule. The phenyl groups bonded to the phosphorus atoms are not shown. In order to locate the symmetry elements of the molecule, consider the following: (*a*) the central atom is an oxygen bridging four coppers (thus μ_4-oxo-), (*b*) there are six chlorines, one for each edge of the tetrahedron, bridging the coppers at the corners (thus hexa-μ-chloro), and (*c*) there are four phosphine oxide molecules, one for each copper (thus the tetrakis phosphine-oxide).

J. A. Bertrand, *Inorganic Chemistry*, **6**, 495 (1967).

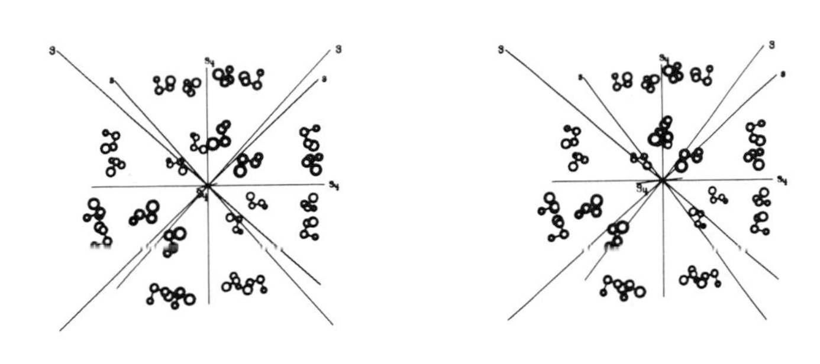

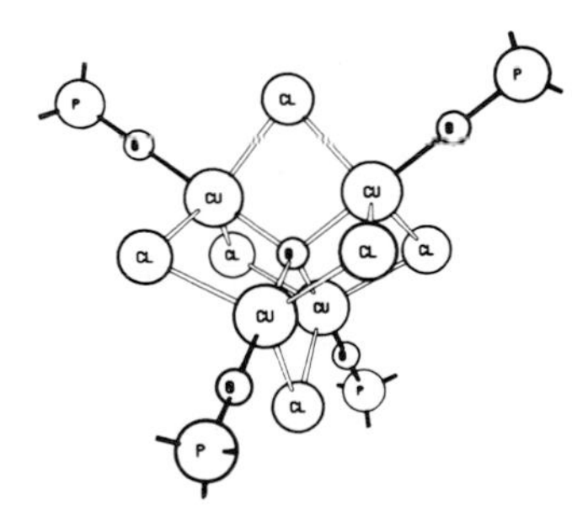
P
CL
CU
CU
CL
CL
CL
CL
CU
CU
CL
P
P

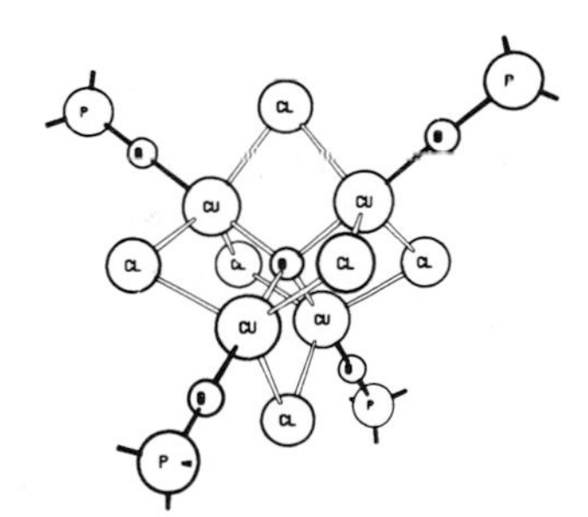
P
CL
CU
CU
CL
CL
CL
CL
CU
CU
CL
P
P

O 432

There are twenty-four symmetry elements. Each of the coordinate axes is a 4-fold axis. The 3-fold axes lie equidistant from the coordinate axes. Finally, in each of the planes defined by two coordinate axes, 2-fold axes lie halfway between the 4-fold axes. All objects are of the same handedness. A molecule with this symmetry would be optically active.

Order of group: 24

Crystallographic symmetry: Cubic

Octamethylcubane
$(CH_3)_8C_8$

This is a possible structure for a **hydrocarbon** of a type that has been synthesized only recently. Carbon atoms lie at the corners of a cube, each forming three C—C bonds on the cube edges and one additional bond. The methyl groups in this hypothetical example are rotated in such a way that no mirror planes would lead to the point group O_h of cubane itself.

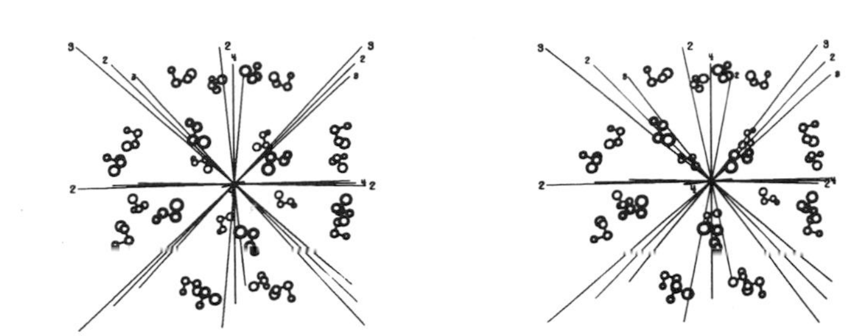

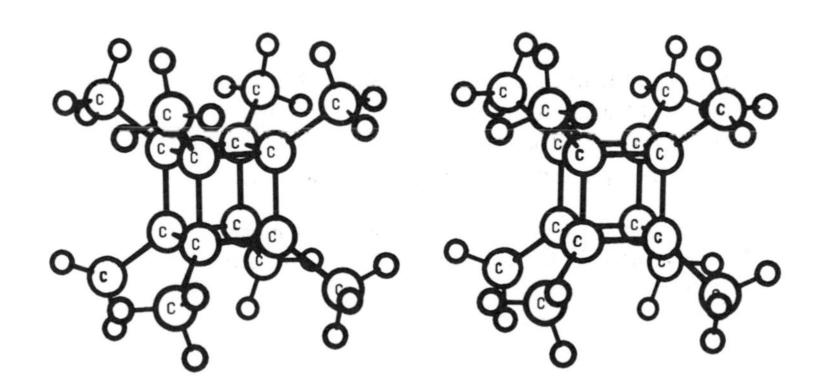

O_h m3m

The same 2-fold, 3-fold, and 4-fold axes of the point group O are contained in point group O_h. There are, in addition, mirror planes that include the coordinate axes by pairs and that also lie in intermediate positions. A center of symmetry is also generated. This point group has the symmetry of the regular octahedron. Objects of each hand are present, and a molecule with O_h symmetry would not be optically active.

Order of group: 48

Crystallographic symmetry: Cubic

Cubane
(C_8H_8)

This beautifully symmetric hydrocarbon was first synthesized in 1964. The 3-fold axes lie along the C—H bonds. The crystal structure in fact contains this 3-fold axis as a symmetry element, although the other symmetry elements of the molecule are not repeated in the crystal.

E. B. Fleischer, *Journal of the American Chemical Society*, **86**, 3889 (1964).
P. Eaton and T. Cole, *Journal of the American Chemical Society*, **86**, 3157 (1964).

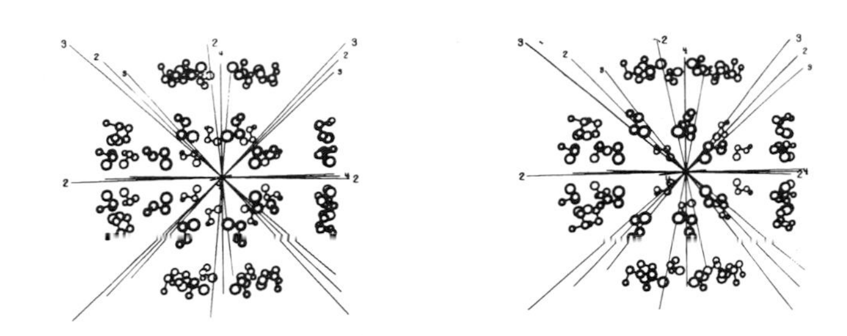

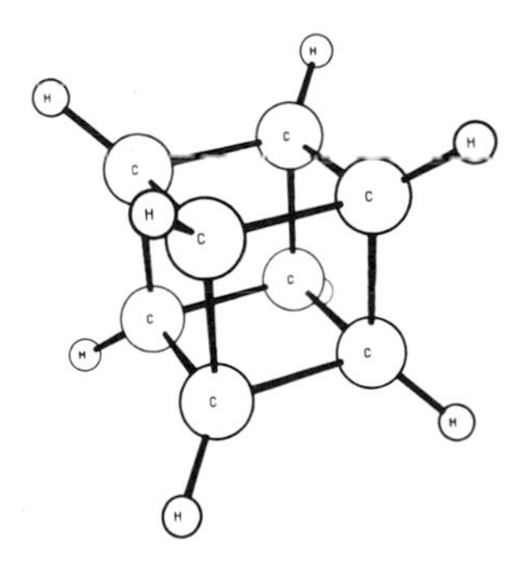
H
C

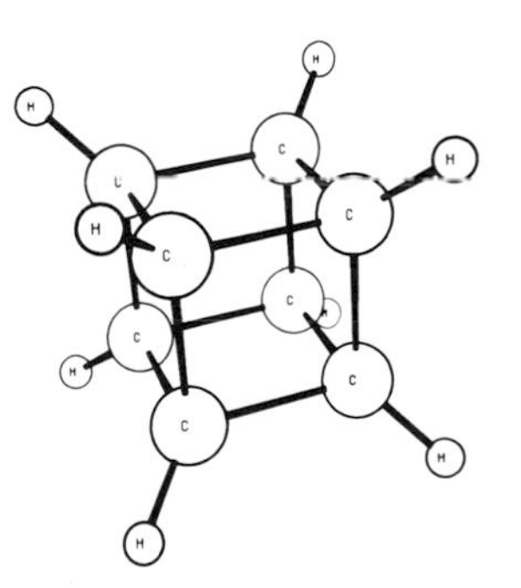
H
C

T_h $m\bar{3}$

This point group is obtained from the point group T by adding mirror planes that include the coordinate axes by pairs. A center of symmetry is generated, too. Objects of both hands are present, and a molecule with this symmetry would not be optically active.

Order of group: 24

Crystallographic symmetry: Cubic

Hexapyridineiron(II) tetrairontridecacarbonylate $[Fe(C_5H_4N)_6][Fe_4(CO)_{13}]$

The cation in this complex (hexapyridine iron(II), which contains six **pyridine** rings around the central iron atom) is one of the few molecular species that belongs to this centrosymmetric cubic point group. The hydrogen atoms on the pyridine rings are not shown.

R. J. Doedens and L. F. Dahl, *Journal of the American Chemical Society*, **88**, 4847 (1966).

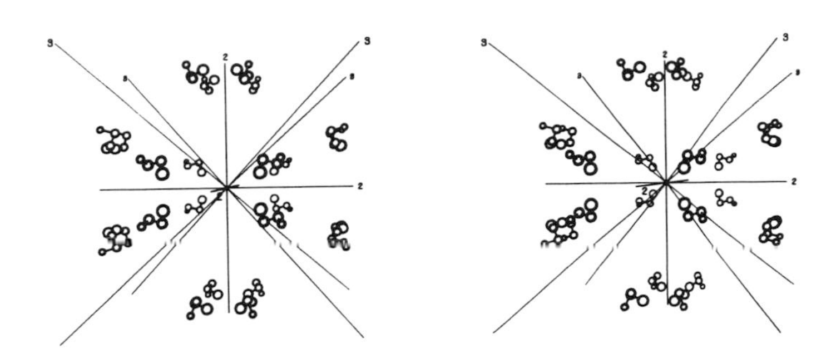

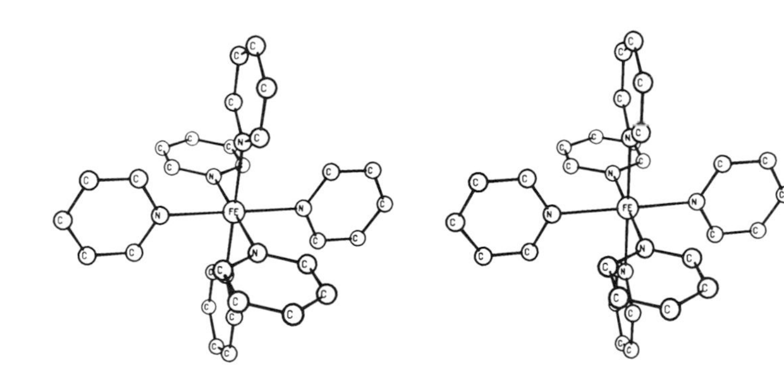
FE
N
C

Problems

1. Give the plane point group symmetries of the following planar molecules.

(*a*) benzene

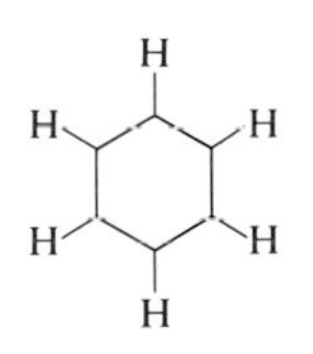

(*b*) napthalene

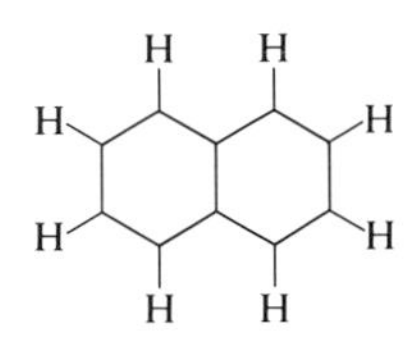

(*c*) the free radical

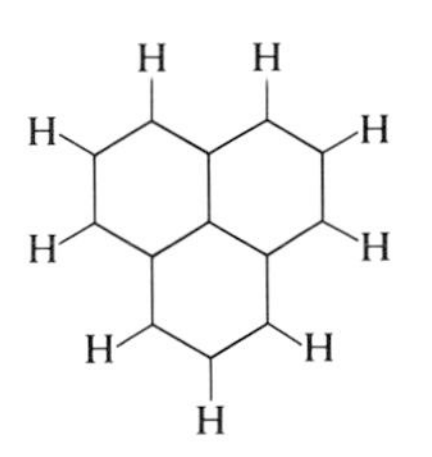

(*d*) 1-chloro-3-bromobenzene

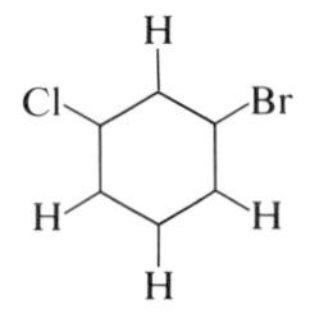

(*e*) 2,6-dichloronapthalene

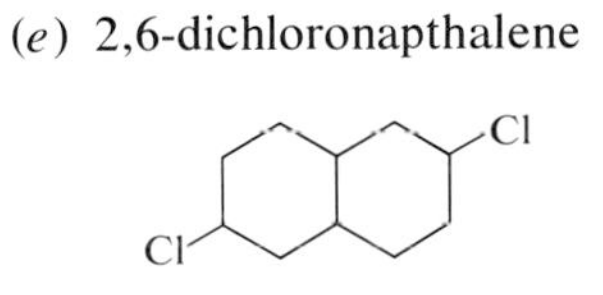

(*f*) 1,4-dichloronapthalene

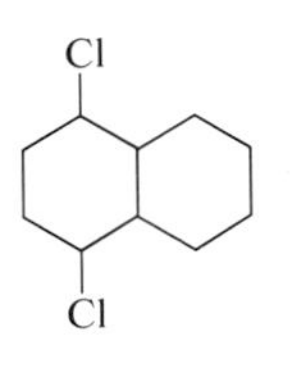

(*g*) the 1,4,7-trichloro derivative of (*c*)

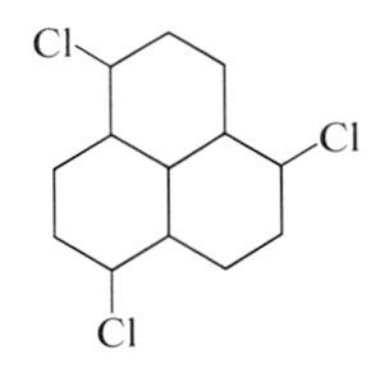

(*h*) cyclobutadiene

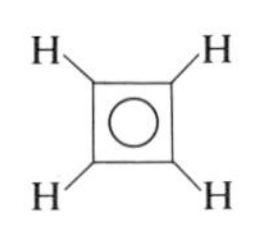

2. The molecule *p*,*p*′-dinitrobiphenyldisulfide crystallizes in a space group that requires the molecule to have a specific symmetry element. What is the symmetry element present here? What is the point group?

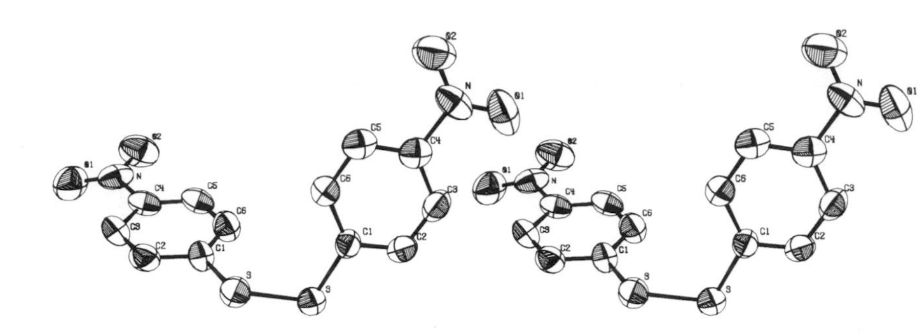

3. The *molecule* of phthalamide. What is the approximate molecular symmetry?

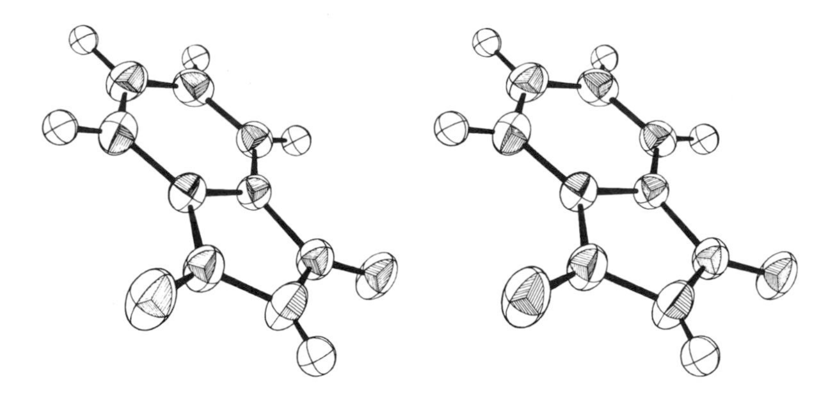

4. A molecule in the crystal structure of $Mn(CO)_5H$, manganese pentacarbonyl hydride. What is the approximate symmetry of the molecule?

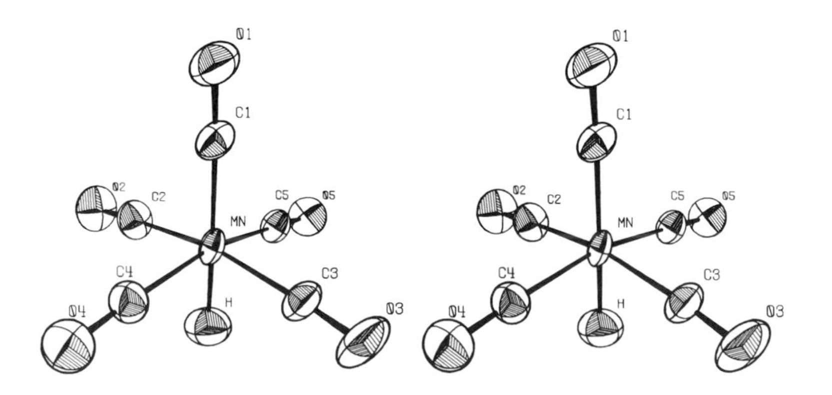

5. What is the approximate (idealized) symmetry of the iron compound shown here? We have presented a nonstereo figure of the molecule (depth is indicated with foreshortened lines) in order to give the reader practice in recreating the third dimension in his mind. The same is true of a few subsequent problems.

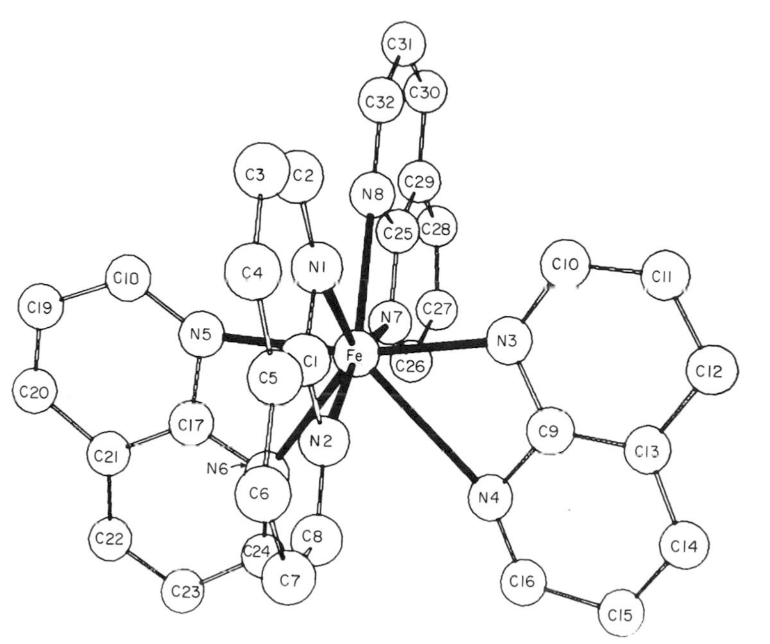

6. Examine only either the left-eye or the right-eye view of the iron compound, bis(cyclopentamethylenedithiocarbamato)iron dicarbonyl, in the following illustration. It is far easier to see the location of a two-fold axis in a stereo view than in a figure that does not present depth vividly. Find the 2-fold axis and give the molecular symmetry.

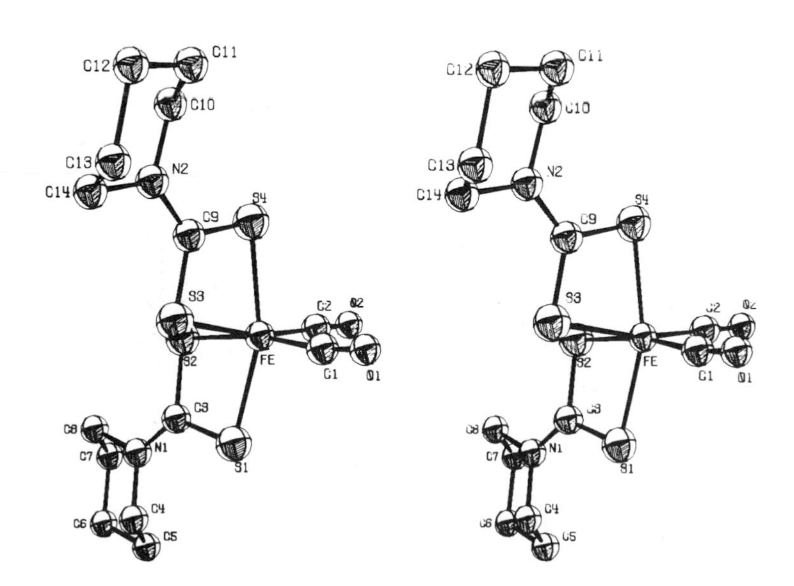

7. What is the approximate point group symmetry of the Brooklyn Bridge or the Golden Gate Bridge?

8. This perchlorate ion (ClO_4^-) ion was found in the same structural study as the iron species illustrated in Problem 5. Disregarding the shapes and orientations of the vibrational ellipsoids, what is the idealized point group?

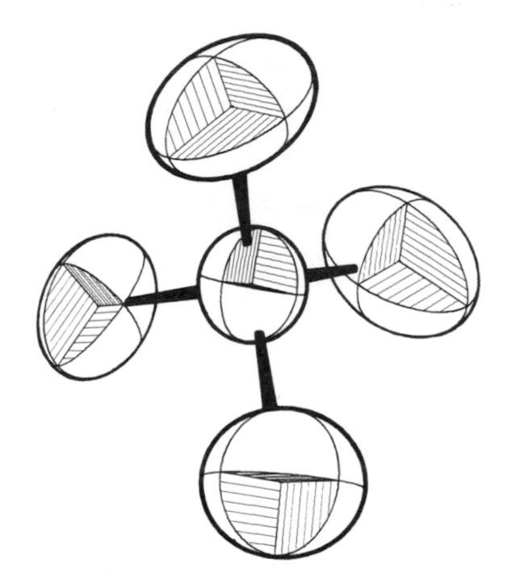

9. This is a tetraphenylarsonium cation $\left((C_6H_5)_4As\right)^+$. What is the point group at the arsenic if (*a*) the benzene rings are idealized to single round balls, and (*b*) all the atoms in the benzene rings are taken into account?

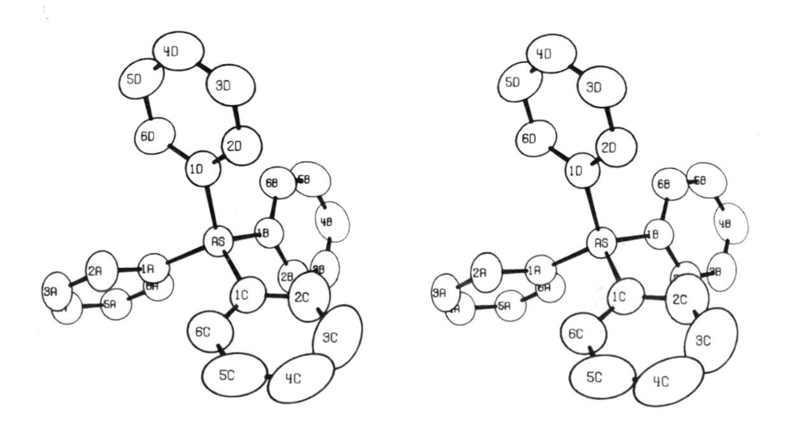

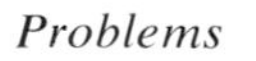

10. The species $CdCl_5^{3-}$ was studied by x-ray diffraction techniques. What is the point group symmetry of this figure if (*a*) account is taken of the canting of the ellipsoids of motion in the basal plane, and (*b*) one disregards the ellipsoids of motion and assumes them to be spheres, instead of ellipsoids?

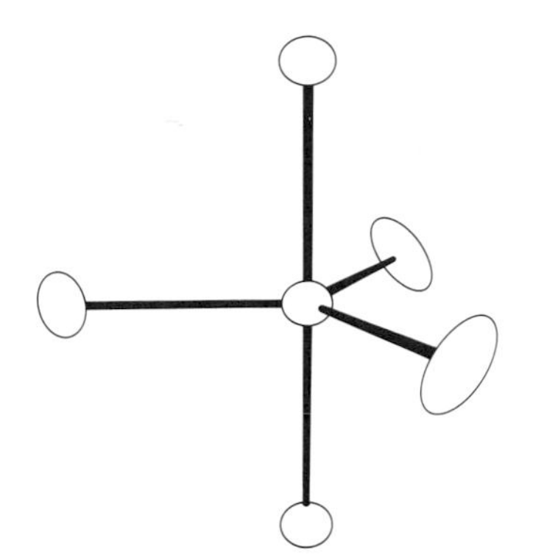

11. A part of the oxygen atom framework of ordinary ice (ice Ih). What is the symmetry of the illustrated figure?

12. A part of the oxygen atom framework in cubic ice (ice Ic). What is the symmetry of the illustrated figure?

13. What is the approximate symmetry of twin brothers facing each other with arms at their sides?

14. Suppose each twin places only his right hand on his brother's left shoulder?

15. This figure depicts the environment of potassium ions in $K_2BaCo(NO_2)_6$. What is the point group at the potassium ions?

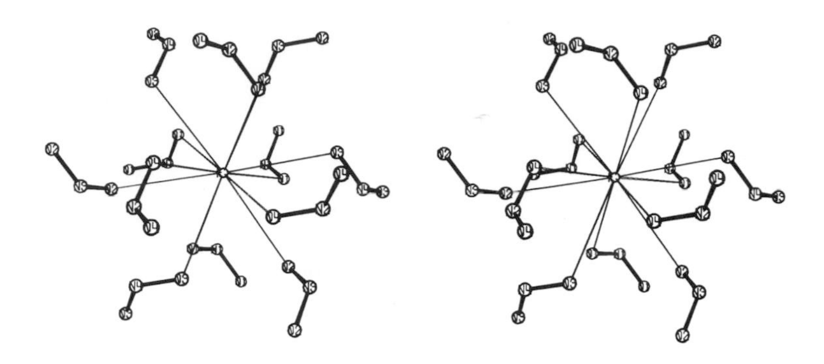

16. In the crystal structure of KH_2PO_4, potassium dihydrogen phosphate, the tetrahedral phosphate ion is linked by hydrogen atoms to other phosphate ions. What is the point group symmetry at the phosphorus atom that lies at the center of the tetrahedron?

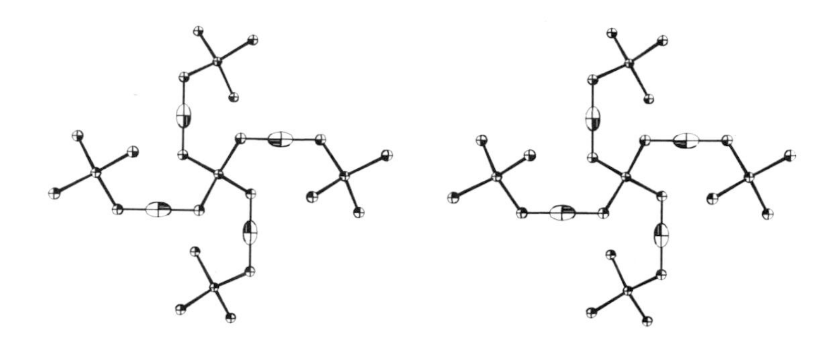

17. The crystal structure of ammonia, NH_3, consists of distinct NH_3 molecules linked together by hydrogen bonds. (*a*) What is the point group symmetry of a single ammonia molecule? (*b*) What is the point group symmetry at the nitrogen atom?

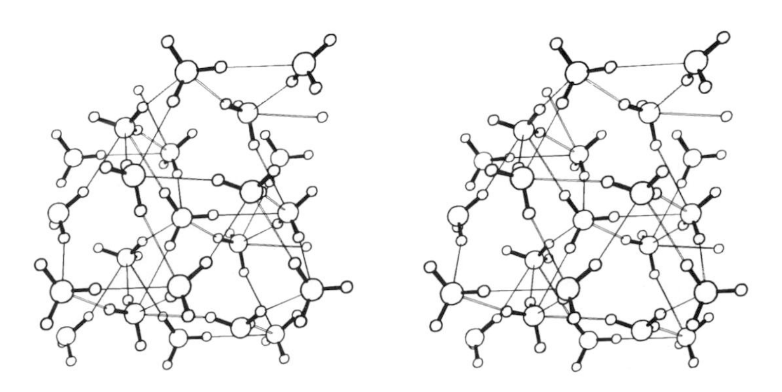

18. This substance is an idealized version of bis(ethylenediamine)$CoCl_2$ in which the five-membered rings defined by the Co—N—C—C—N fragments should be assumed to be planar. Under those conditions, what is the point group of this molecule?

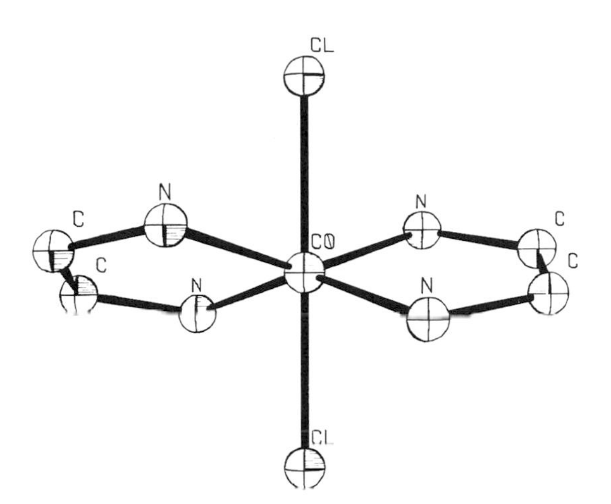

19. Molecules of bis(perfluoromethyldithiolato)iron triphenylarsine are so aligned in the solid state that pairs of them interact as shown in the figure. (*a*) What is the point group of the pair? (*b*) What is the approximate (idealized) point group of each half?

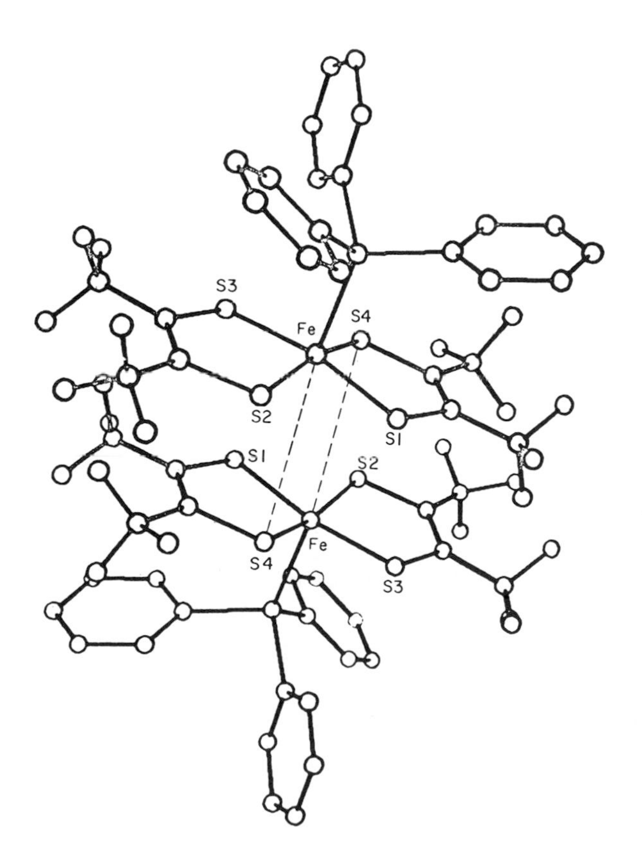

20. The alloy of rhodium and beryllium, $RhBe_{6.6}$, illustrated on page 16 contains a number of interesting polyhedra of high symmetry. Give the symmetry operations for those figures and thus determine the point group for the species shown in them.

21. This figure shows one of the potassium atoms in a copper salt having composition $K_3Cu(NO_2)_5$. What is the point group at the potassium atom?

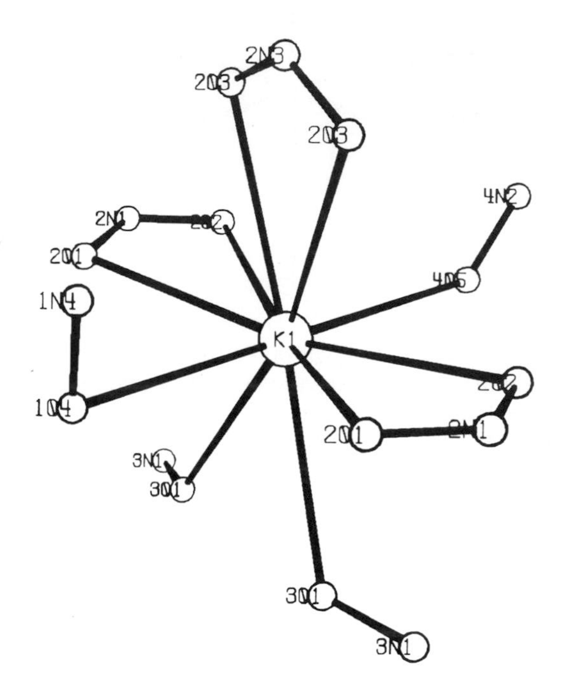

22. This is one of the copper species present in $K_3Cu(NO_2)_5$. What is the point group at the copper?

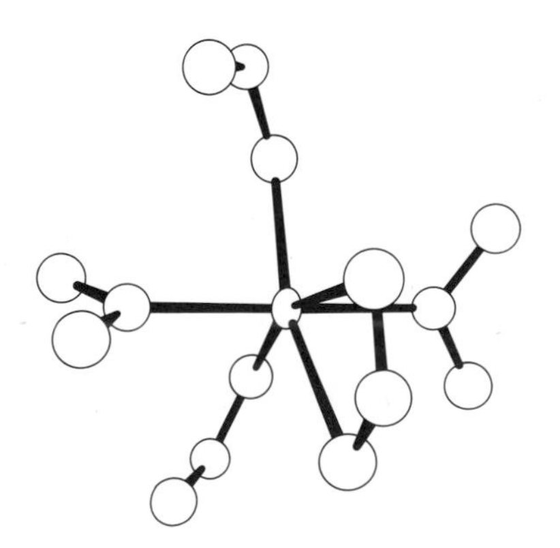

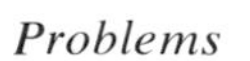

23. This is another copper species present in $K_3Cu(NO_2)_5$; this time the picture is in stereo. Is it easier to appreciate the symmetry elements present? What is the point group? If the numbers labelling the atoms are ignored, what is the point group?

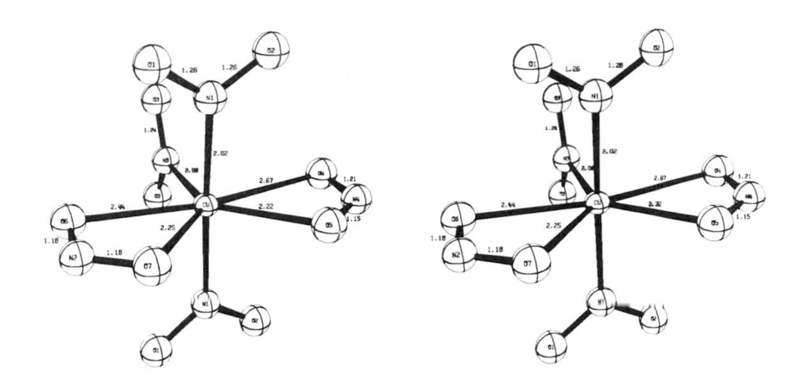

24. The crystal structure of trichloroacetic acid trimer. What is the symmetry of an individual dimer?

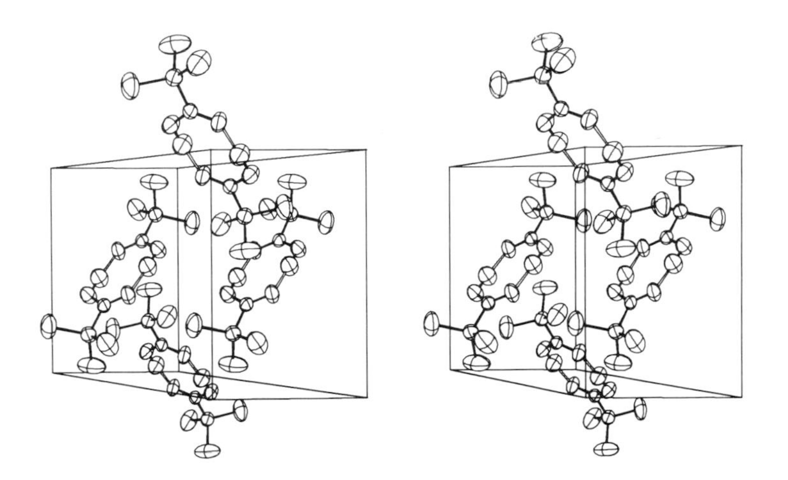

25. Surroundings of the iodide ion in PH_4I-phosphonium iodide. What is the point group symmetry at the iodide ion?

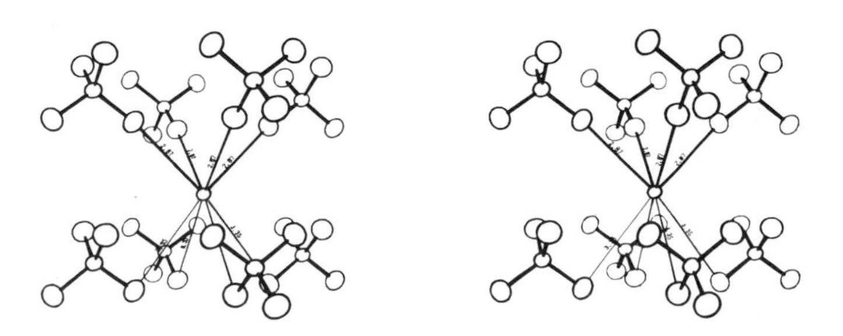

26. The structure of diamond (one unit cell, and eight unit cells). What is the point group symmetry, viewing a carbon atom as the symmetry point?

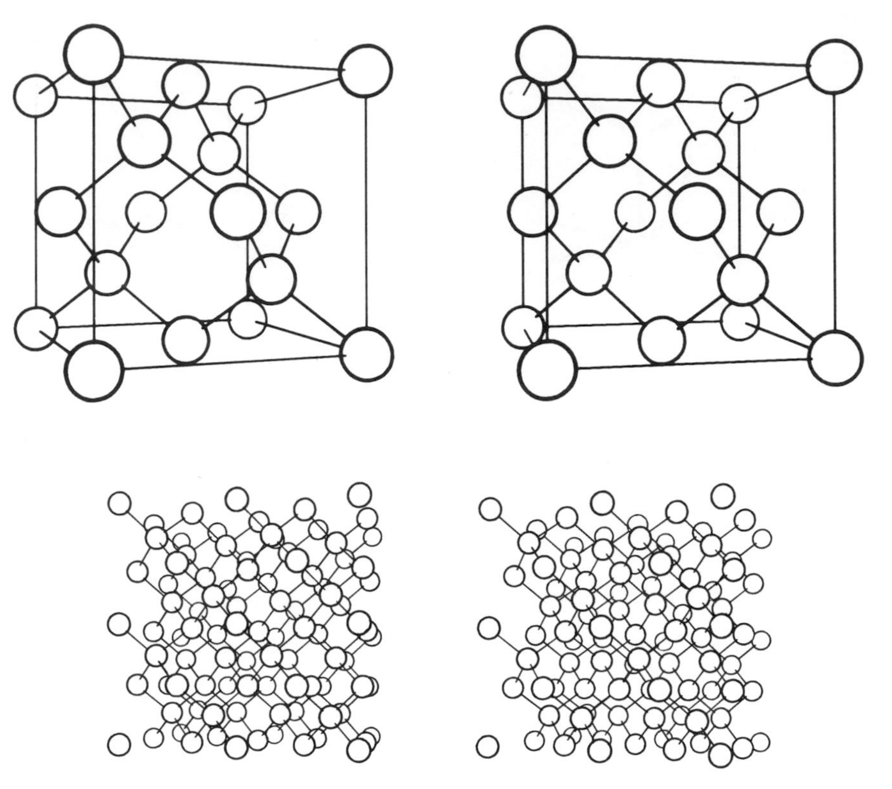

27. What is the point group symmetry of diamond, taking the midpoint of the C—C bond as the origin?

28. The environment of Li in $LiClO_4 \cdot 3H_2O$, in which lithium is surrounded by six water molecules. What is the point group at the central atom?

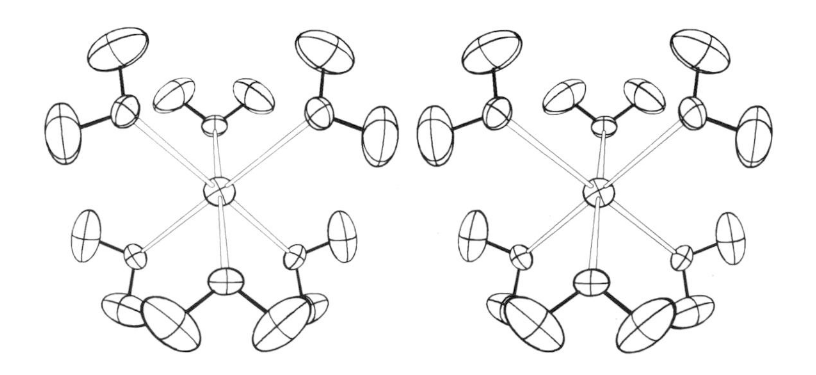

29. The perchlorate ion in $LiClO_4 \cdot 3H_2O$ has one oxygen (labelled O1) whose nearest neighbors are three waters, as shown. What is the point group of the aggregate?

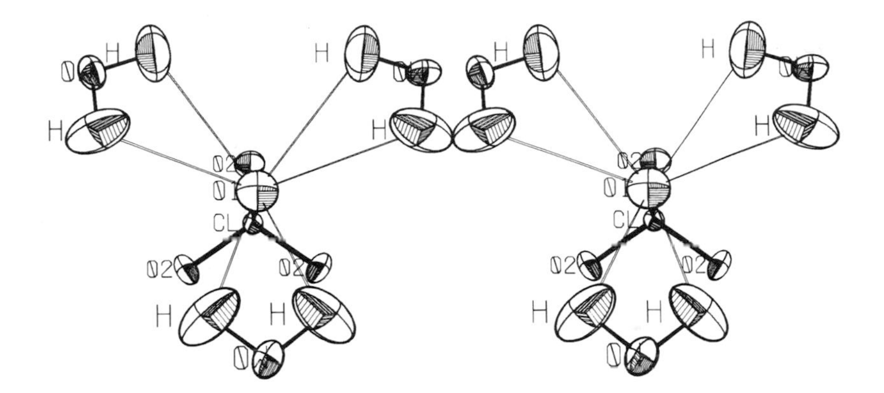

30. The crystal structure of white tin is shown in the figure. What is the point group symmetry at the central tin atom?

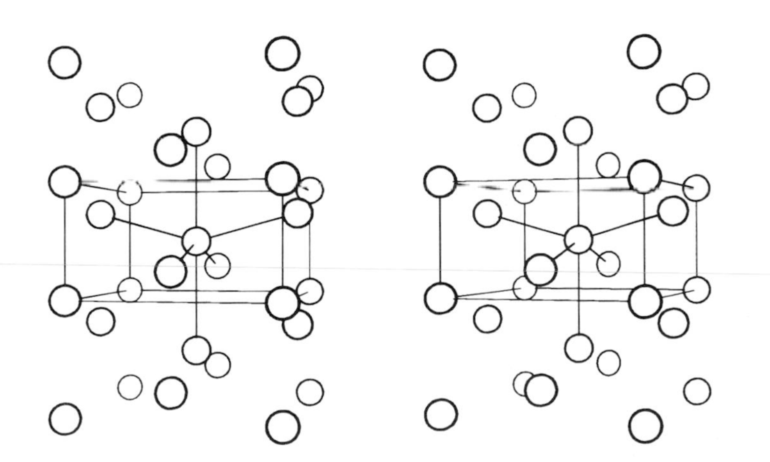

31. One of the simplest crystal structures is that composed of a close packing of spherical atoms. Illustrated here is the hexagonal close-packed structure. What is the point group symmetry at the atom positions?

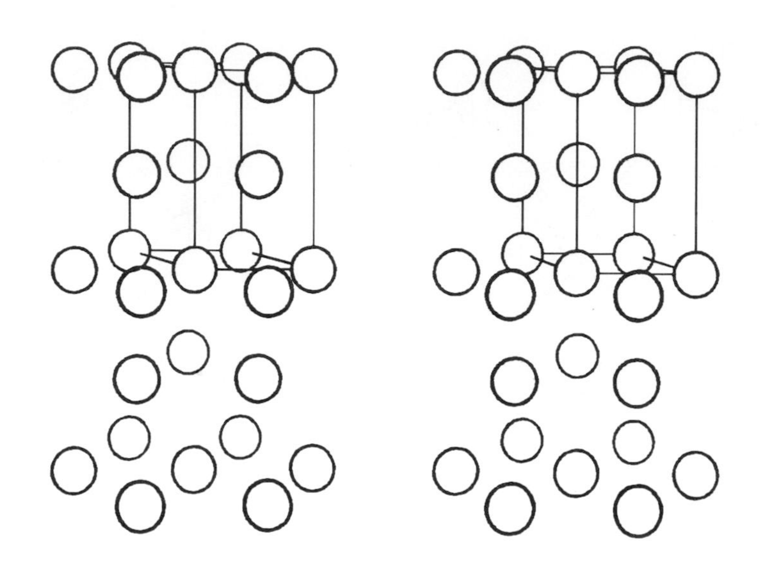

32. An alternate packing of spheres is cubic closest-packing. What is the point group symmetry at the atomic site?

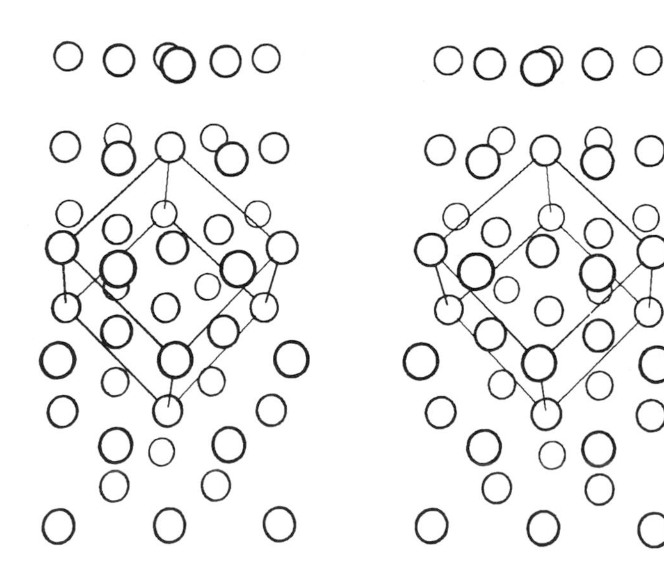

33. This is a special exercise. What is the operation that transforms the molecules into each other in the spirals shown? The operation in question involves a rotation followed by a translation along the length of the spiral. Such operations are called screw axes.

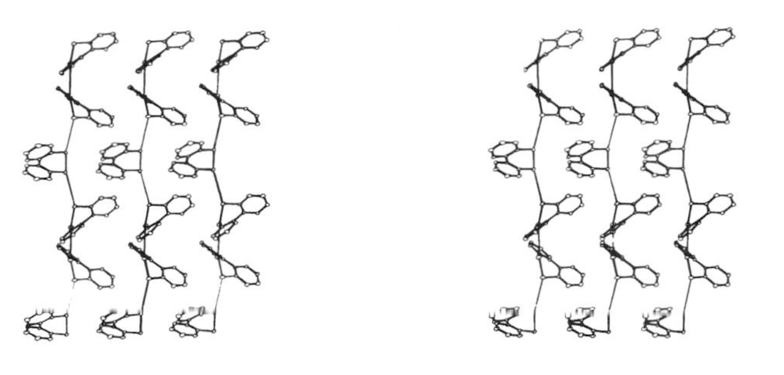

34. What is the plane point group of (*a*) the total stamp, (*b*) the design only, without the lettering, and (*c*) the design if the two fish are omitted?

Answers to Problems

1. (*a*) 6*mm*, (*b*) 2*mm*, (*c*) 3*m*, (*d*) 1, (*e*) 2, (*f*) *m*, (*g*) 3, (*h*) 4*mm*
2. A 2-fold axis. $C_2 \equiv 2$
3. $C_{2v} \equiv 2mm$
4. $C_{4v} \equiv 4m$
5. $D_{2d} \equiv \bar{4}2m$
6. There is a 2-fold axis passing halfway between the two carbonyl (CO) groups, and the chair-shaped six-membered rings are correctly oriented to be flipped into each other by the same 2-fold axis operation. The symmetry is $C_2 \equiv 2$.
7. $C_{2v} \equiv 2m$
8. $T_d \equiv \bar{4}3m$
9. $T_d \equiv \bar{4}3m$; $C_1 \equiv 1$
10. (*a*) $D_3 \equiv 32$, (*b*) $D_{3h} \equiv \bar{6}m2$
11. $D_{3h} \equiv 3/mm$
12. $D_{3d} \equiv \bar{3}m$
13. 2*mm*
14. 2
15. $T_d \equiv \bar{4}3m$
16. $S_4 \equiv \bar{4}$
17. (*a*) $C_{3v} \equiv 3m$; (*b*) $C_3 \equiv 3$
18. $D_{2h} \equiv mmm$
19. (*a*) $S_2 \equiv C_i \equiv \bar{1}$; (*b*) $C_{1h} \equiv C_s \equiv m$

20. (*a*) $C_{3v} \equiv 3m$, (*b*) $D_{3h} \equiv \bar{6}m2$, (*c*) $C_{3v} \equiv 3m$, (*d*) $C_{3v} \equiv 3m$. Note the polyhedron at the bottom of page 16 would have perfect icosahedral symmetry if all the atoms were identical.

21. $C_{1h} \equiv C_s \equiv m$
22. $C_{1h} \equiv C_s \equiv m$
23. $C_{1h} \equiv C_s \equiv m$; $C_{2v} \equiv 2mm$
24. $C_2 \equiv 2$
25. $D_{2d} \equiv \bar{4}2m$
26. $T_d \equiv \bar{4}3m$
27. $D_{3d} \equiv \bar{3}m$
28. $C_{3v} \equiv 3m$
29. $C_3 \equiv 3$
30. $D_{2d} \equiv \bar{4}2m$
31. $D_{3h} \equiv \bar{6}m2$
32. $T_d \equiv \bar{4}3m$
33. The operation is a 3-fold screw axis, expressed as a 3_2 screw axis.
34. (*a*) 1, (*b*) 2, (*c*) 2*mm*

Glossary

A number of chemical terms used in the textual material may not be familiar to all students. We present here definitions of some of these (indicated with bold face type in the text). Further discussion will be found in first-year chemistry texts, such as L. Pauling's *General Chemistry*, Third Edition, San Francisco: W. H. Freeman, 1970. We make no attempt to be completely general or exhaustive in the brief definitions here.

Acetylacetonate. A derivative of acetylacetone. In organic chemistry, the grouping CH_3—CO— is an acetyl group. Acetone is CH_3—CO—CH_3, and an acetonate contains the radical —CH_2—CO—CH_3. Thus acetylacetone is CH_3—CO—CH_2—CO—CH_3. The acetylacetonate radical is obtained by abstraction of one of the central —CH_2— hydrogens. Other derivatives of acetylacetone are obtained by substituting larger groups for the hydrogens at the methyl (CH_3) or methylene (CH_2) groups. For instance, if $(CH_3)_3C$ groups replace the methyl groups, the substance is known as dipyvaloylmethane or hexamethylacetylacetone.

Acid. In the classical sense, an acid is a compound HA which, when dissolved in water, dissociates into a hydrogen ion, H^+, and a negatively charged species, A^-, called the anion. Acetic acid, for instance, dissociates into a hydrogen ion and an acetate anion. For a more generalized treatment, the reader is referred to Pauling's *General Chemistry*.

Adduct. When two substances, each capable of independent existence, combine to form a single new product, the product is an adduct. In human terms, a marriage.

Amine. An amine is a chemical compound consisting of a central nitrogen atom to which three radicals are attached. If all three radicals are hydrogen atoms for instance, we have the amine NH_3 which we know as ammonia. If we have three methyl groups instead, the compound is known as trimethylamine, etc.

Amino acid. In organic chemistry a carboxylic acid is a substance that can be described by writing R—C(O)OH where R is any configuration of other atoms. If R is a methyl (CH_3) group the acid is acetic acid, or vinegar. If R contains an amino (—NH_2) group the resulting species is called an amino acid. For example NH_2—CH_2—C(O)OH is the amino acid glycine. (*See also* Acid.)

Antiprism. A prism is a solid with a set of edges all parallel to a single axis (the prism axis) and terminated by a pair of congruent parallel polygonal faces which intersect the prism axis. In an antiprism, the two terminating faces are rotated with respect to each other, thus doubling the number of faces parallel to the prism axis. The trigonal antiprism is thus an octahedron; the tetragonal antiprism is a dodecahedron.

Aromatic. Early chemists noted that a class of compounds was characterized by having distinctive, generally pleasant, smells; thus, they referred to them by the qualitative term "aromatic." Many such compounds have since been shown to contain electrons that are very mobile, or delocalized over a large portion of the molecule. Compounds of this special class are thus called aromatic. Benzene and naphthalene (moth balls) are typical of these substances.

Beta-carbon. If an atom, such as a carbon, is one, two, or three atoms away from a group that is used as a marker or starting point of counting, such a carbon is said to be an α- (alpha), β- (beta), or γ- (gamma) carbon with respect to the marker. (*As an example, see* Beta-diketone.)

Beta-diketone. A ketone is a species containing the grouping —CO—, and a diketone contains two of them. The relative positions of the carbonyl (CO) groups to one another is determined by the prefix used before the word "diketone." In general, diketones can be written as R—CO—$(CH_2)_n$—CO—R, where R is a radical, such as H, CH_3, etc. If n=0 and there is nothing between the two carbonyl groups, the species is known as an alpha-diketone; if n=1, the species is a beta-diketone, etc. (*See* Acetylacetonate.)

Bidentate. Dentate means having teeth, the implication being that a dentate species can hold onto or sink its teeth into something else. A species that can hold onto another by two sets of its "teeth" is a bidentate species. Beta-diketones can, for example, bind a metal atom with the two carbonyl "teeth."

Bis. If a species appears n times at chemically equivalent positions, the overall compound is said to be a bis, tris, tetrakis, pentakis, hexakis, heptakis, octakis, etc., compound if $n = 2, 3, 4, 5, 6, 7, 8$, etc. Bis(cyclopentadienyl)iron, for instance, is an iron compound that contains two cyclopentadienyl anions ($C_5H_5^-$). It is commonly known as ferrocene.

Bridge bond. When two species, A and B, are linked by a third one, C, usually a single atom, the species C is called a bridge and the bonds formed are called bridge bonds. The nomenclature used to denote a bridge bond is the Greek letter μ (mu) as a prefix of the name of the bridging species. For example, if two iron atoms are linked by a chlorine bridge, Fe—Cl—Fe, the fragment or species would then be called a μ-chlorodiiron fragment.

Cation. Any positively charged species.

Chirality. From the Greek word *cheiro*, meaning hand. Thus, chirality means handedness, as in left or right handedness.

Chlorophyll. The green pigment of the leaves of plants. It converts carbon dioxide and water, under the influence of sunlight, into sugars. The central portion of the molecule contains a magnesium porphine. (*See* Porphine.)

Cluster compound. A specialized name for those compounds having metal atoms in discrete clusters containing, usually, four or more metal atoms bonded to one another. In a sense, they can be viewed as discrete little chunks of metals.

Complex; Complex ion. The term complex or complex ion was first used in the latter part of the nineteenth and early part of the twentieth centuries to describe ions or compounds whose nature was too "complex" to be rationalized with the ideas then available to explain bonding. The problem, as it existed then, can be illustrated by the following example: Take a neutral, stable compound such as cobalt trichloride, $CoCl_3$, containing a Co^{3+} cation and three Cl^- anions. The chemists of those days considered it to be a "saturated" compound, that is, one whose total combining capacity is satisfied since it is well defined and stable. By placing this substance in aqueous solutions of another saturated species, ammonia (NH_3), chemists isolated new, stable compounds having composition $Co(NH_3)_xCl_3$, with x ranging from 3 to 6. Obviously the idea that $CoCl_3$ was a "saturated" compound was only correct with respect to chloride but not with respect to ammonia. It was also found that two species with identical x values can have different physical properties. Thus, composition alone could not account for the complexity of such compounds; and it became clear that the relative spatial distribution of the bonding species around a metal ion was the answer to the riddle. The nature of complex compounds was explained by Alfred Werner, a Swiss chemist who won the Nobel Prize in 1913 for this great achievement. Werner proposed an octahedral array of six species around the metal as a model capable of answering for all the physical properties of the compounds in question. The theory was contrary to the established facts of carbon chemistry. The reader can gauge the amount of opposition to Werner's theory by the fact that he proposed it in 1893 and at the time of the Nobel award he was still actively trying to provide answers to the attacks of his critics! (*See* Ion.)

Coordinated. The chemists who coined the term complex ion realized early that for each specific metal ion there always was a fixed number of ligands of a given kind surrounding the metal. Thus, they spoke of the number of ligands a metal could fix or "coordinate" to itself. They also, therefore, spoke of coordination or complex compounds as being synonymous terms. (*See* Complex ion.)

Delocalized bond. In a single bond two electrons are located between the two atoms being bonded. A pair of atoms held by a double bond contain, in addition to the two electrons of the single bond, two electrons located in the space above and below the single bond. Certain substances contain single and double bonds in such a sequence that more than one equivalent arrangement is possible. A good example is benzene:

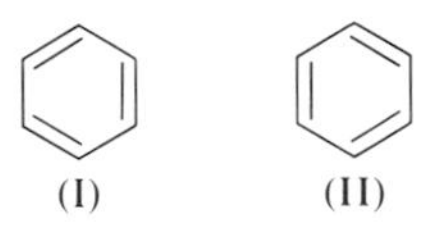

(I) (II)

The molecule represented in II is chemically identical to molecule I. Because the electrons in these molecules are delocalized in space, the bonds they form are known as delocalized bonds.

Dihydrate. A species containing two water molecules. (*See* Hydrate.)

Dimeric. A dimer or dimeric species is an entity composed of two identical halves.

Dipivaloylmethanide. A derivative of dipivaloylmethane, which is in turn a beta-diketone. (*See* Beta-diketone.)

Dodecahedron. A solid figure bounded by twelve planes.

Eclipsed. If, along the line of sight, the atoms of a particular group of a molecule lie directly in line with those of a similar group, the relative conformation of those two groups is said to be eclipsed. If they are so oriented instead that the atoms of one are aligned halfway between those of the other group, the conformation is said to be staggered.

Enzymes. Enzymes are large molecules whose biological function is to act as catalysts for chemical reactions necessary to maintain living processes functioning efficiently. Most enzymes are very specific in their function in that they operate on one, and only one, chemical substance or reaction. For instance, saliva contains an enzyme (amylase) whose function is to break down starch from the food we eat. (*See* Protein.)

Five-coordinate. Describing a species in which a central atom is surrounded by five species known as ligands. (*See* Ligand; Complex ion; Coordinated.)

Hexamer. A compound composed of six identical parts. (*See* Polymer.)

Hemoglobin. A protein found in blood whose function is to transport oxygen molecules through the body. The portion of the molecule where the oxygen is bound is an iron porphine. (*See* Porphines.)

Hexahydrate. Containing six water molecules. (*See* Hydrate.)

Hydrate. Containing water. The word is derived from the Greek word for water and whenever a species contains n water molecules it is said to be monohydrated, dihydrated, trihydrated, etc., if it contains one, two, or three molecules of water. The term is used indiscriminantly in the sense that no distinction is made between water molecules bound to a species, as in a coordination or complex compound, and water molecules that merely fill interstices or cavities in a crystalline lattice.

Hydrocarbon. A compound containing only hydrogen and carbon. Examples are methane (CH_4), benzene (C_6H_6), acetylene (C_2H_2), and octane (C_8H_{18}).

Icosahedron. A solid bound by twenty planes or faces.

Ion. A chemical compound that is not electrically neutral. If there is an excess of electrons, we have a negatively charged species—an anion. If there is a deficiency of electrons, we have a positively charged species—a cation.

Ionic crystal. Solids containing an orderly array of charged species. The positively charged ones, cations, and negatively charged ones, anions, exist in numbers such that the overall charge of the solid is zero.

Ligand. From the Latin *ligare*, to bind; for example, a ligament is something that ties one part to another. In inorganic chemistry a ligand is a species capable of binding a metal ion to itself.

Octahedral. A regular octahedron (a polyhedron with eight faces) is described by a set of six points, equidistant from a common origin along three perpendicular lines. In modern inorganic chemistry, octahedral refers to any complex species having approximately this coordination of six ligands around a central atom.

Optically active. A bulk substance capable of rotating the plane of polarization of an electromagnetic wave is said to be optically active. For simple and clear discussions of the relationship between symmetry, molecular configuration and optical activity, the reader is referred to an article by J. L. Carlos, Jr., *Journal of Chemical Education*, **45**, 248 (1968) and also to Pauling's *General Chemistry*.

Pentacoordinated. Another term for **five-coordinate**.

Pentagonal bipyramid. A polyhedron consisting of two pyramids having a common five-sided base. There are 10 faces, 7 vertices, and 15 edges.

Peptide. A peptide linkage is one derived from the reaction in which an organic acid (R—COOH) and an amine (R_2NH) lose a water molecule and in the process produce a species R—CO—NR_2. Organic chemists call this compound an amide. If the reaction is between amino acids, the resulting compound is called a peptide, which always still contains an amine group and a carboxylic acid group and is thus capable of undergoing polymerization.

Pi-bond. A chemical bond is formed when two atoms share two electrons. If the two atoms share more than two electrons, a multiple bond exists. In such a situation, it is convenient to describe two types of bonds. First, the single bond formed by the sharing of the first two electrons is called a σ- (sigma-)bond. The electron density lies approximately along the interatomic line and has cylindrical symmetry. Additional bonds place electron density above and below the interatomic line. These are called π- (pi-)bonds. They occur in ethylene $CH_2{=}CH_2$ and in benzene, where six carbon atoms share 18 electrons in six σ-bonds, and three delocalized π-bonds. Electrons in π-bonds are frequently available for binding to other species.

Polyhedron. A solid figure bound by a number of planes.

Polymer. A compound formed by linking large numbers (thus the poly- part of the word) of small units that repeat at regular intervals. If the unit appears n times, the species is called a monomer, dimer, trimer, tetramer, pentamer, hexamer, . . . , polymer for n = 1,2,3,4,5,6, . . . , a large number.

Porphine. A molecule of carbon, hydrogen, and nitrogen having the following chemical structure. (Only one of the possible assignments of single and double bonds is shown.)

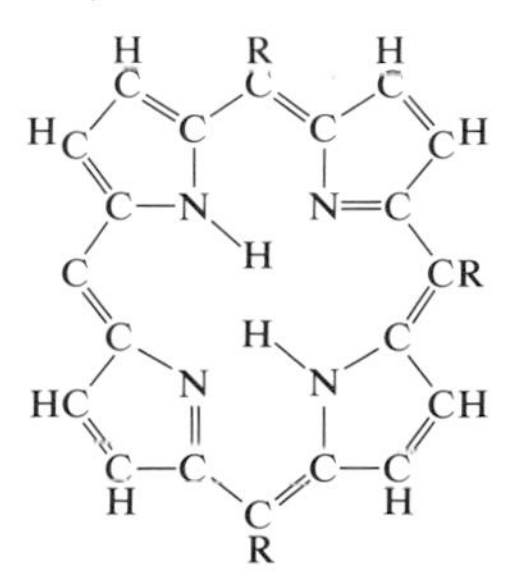

The molecule is completely flat and it is **aromatic**. If R = H, the molecule is porphine; if R = phenyl (C_6H_5) the molecule is known as tetraphenylporphine. If a metal such as iron or magnesium is inserted between the four nitrogen atoms of the five-membered rings, replacing the two hydrogen atoms, the species is known as an iron or magnesium porphine. These are very important species which occur in biologically active molecules such as hemoglobin and chlorophyll.

Protein. A complex compound formed by **polymerizing** α-amino acids through the formation of **peptide** linkages. These high molecular weight polymers usually have molecular weights exceeding 30,000 atomic weight units per molecule and are essential components of all living systems. (*See* Amino acid.)

Pyridine. An **aromatic** compound C_5NH_5, similar to benzene, but with one C—H fragment replaced by N.

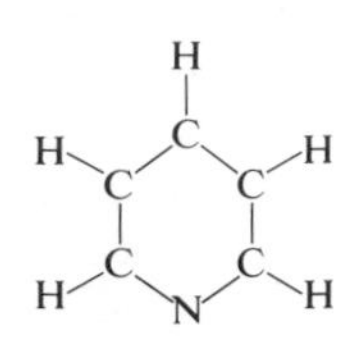

Quantum mechanics. The mathematical physical theory used to describe the properties of atoms and molecules. So named because systems on the atomic size scale can only have energies that differ by specific small amounts—or quanta. In classical mechanics a system can have any energy.

Radical; Free radical. A fragment of a molecule, usually called a radical, is a fragment derived from the rupture of a bond, in which process both fractions carry with them one of the two electrons forming the bond being broken. Thus each of the two fragments ends up having an odd number of electrons. These fragments are called "free radicals" because of the free or uncompensated electron each contains.

Sandwich compound. A compound in which the bread is a pair of flat, **aromatic** rings such as benzene and the hot pastrami is a metal ion.

Steric energy. When two bulky groups are adjacent to each other in a molecule it is possible that their bulk is such that they physically interfere with one another. Often compromises are made in such molecules, and groups acquire configurations that are distortions of what they normally prefer to have while minimizing the steric contacts between the interfering groups. The extra energy the molecule acquires in this process is called the steric energy.

Stoichiometry. The proportions in which elements combine to form chemical compounds. In chemical reactions it also means the number of molecules of A needed to react with B in such a way that mass balance is achieved. The stoichiometry of a compound is simply its molecular formula, with no regard to how the atoms are connected.

Tetrahedral. An array of four particles is said to be tetrahedral if the four particles occupy every other corner of a cube.

Tetragonal antiprism. A solid bound at opposite ends by two squares whose corners are staggered in such a way that in order to superimpose them, one has to be rotated by 45° with respect to the other.

Tetragonal pyramid. A solid bound by a square basal plane and four identical triangular faces coming to a common point or apex.

Trigonal prism. A solid bound by two parallel triangular faces and three rectangular faces each joining pairs of edges of the two triangles.

Trimer. A species composed of three identical smaller fragments. (*See* Polymer.)

Tris. When the same species occurs three times at chemically equivalent positions. (*See* Bis.)

A Brief Bibliography

The student may find the following books helpful for further exploration into the world of molecular symmetry.

F. A. Cotton, *Chemical Applications of Group Theory*, 2nd Edition, New York: Wiley-Interscience, 1971.

Elegant presentation of the mathematical properties of symmetry groups with application to the theory of molecular structure. Lucid and practical, but best suited for the advanced undergraduate or graduate student with some knowledge of theoretical molecular structure.

M. C. Escher, *The Graphic Work of M. C. Escher*, New York: Meredith Press, 1967.

A world of phantasy, symmetry, and antisymmetry developed in a series of remarkable lithographs and woodcuts.

N. F. M. Henry and K. Lonsdale, Eds., *International Tables for X-ray Crystallography, Volume 1: Symmetry Tables*. Birmingham: The Kynoch Press, 1952.

Beautiful presentation of the 230 crystallographic space groups, as well as other practical information about symmetry. First-class drawings, and typography worthy of a prize.

R. M. Hochstrasser, *Molecular Aspects of Symmetry*, New York: W. A. Benjamin, 1966.

A mathematically oriented exposition of the theory of symmetry groups, well-done, with applications to problems in molecular structure . . . "intended to be simple enough for students in certain excellent departments." Best suited for students with some experience in molecular quantum mechanics. Includes character tables for many common symmetry groups.

V. A. Koptsik, *Shubnikov Groups*, Moscow: Moscow University Press, 1966.

A tabulation of the 1651 two-color groups in three dimensions. Includes interesting two-color drawings for each of the groups. Hard to obtain, but well worth a study for those interested in one of the more esoteric branches of symmetry.

W. N. Lipscomb, *Boron Hydrides*, New York: W. A. Benjamin, 1963.

Although not specifically concerned with symmetry, this monograph and the monograph *Polyhedral Boranes* listed below nevertheless present some beautiful examples of symmetry at work in a remarkable class of chemical compounds.

C. H. MacGillavry, *Symmetry Aspects of the Work of M. C. Escher*, Utrecht: A. Oosthoek, 1965.

Some of the best of Escher's plane drawings analyzed according to their point group symmetries.

E. L. Muetterties and W. H. Knoth, *Polyhedral Boranes*, New York: Marcel Dekker, 1969.

L. Pauling and R. Hayward, *The Architecture of Molecules*, San Francisco: W. H. Freeman, 1964.

The beauty of molecules and crystals handsomely illustrated. A must for the chemistry student with any aesthetic sensibilities.

P. Stapp, Dir., *Symmetry*, New York: Contemporary Films–McGraw-Hill, 1967.

An animated, educational film that has won international prizes. A sheer delight for all ages; it was designed by a professional film maker in collaboration with three physicists: Judith Bregman, Richard Davisson, and Alan Holden.

H. Weyl, *Symmetry*, Princeton: Princeton University Press, 1952.

Four illuminating and delightful lectures on, symmetry in art, architecture, natural science, and mathematics. Essential reading for lovers of symmetry.

Index to Symmetry Groups

Index to Chemical Compounds